GREEN GOSPEL

Forested Natural Bridge in Sewanee, Tennessee
Illustrations by Mary Priestley

GREEN GOSPEL

Foundations of Ecotheology

JOHN GATTA

Unless otherwise noted, illustrations by Mary Priestley.

Church Publishing Incorporated
19 East 34th Street
New York, NY 10016
www.churchpublishing.org

Cover image by NASA, 1968
Cover design and typeset by Nord Compo

Library of Congress Cataloging-in-Publication Data

Names: Gatta, John, 1946– author.
Title: Green gospel : foundations of ecotheology / John Gatta.
Description: New York, NY : Church Publishing, [2024]
Identifiers: LCCN 2023038098 | ISBN 9781640656628 (pbk) | ISBN 9781640656635 (ebook)
Subjects: LCSH: Ecotheology.
Classification: LCC BT695.5 .G375 2024 | DDC 261.8/8—dc23/eng/20231102
LC record available at https://lccn.loc.gov/2023038098

For Arnold Benz and Ruth Wiesenberg Benz,
true friends through decades and across continents

Contents

Preface

I am scarcely the first to sense that the current climate emergency, as it's rightly been called, not only poses a grave physical challenge to life on earth as we know it but also signifies a crisis of spirit. And while each of the world's major faith traditions contributes something to our understanding of this peril, the reflections that follow focus on the gifts for healing both the earth and ourselves that I take to be offered in abundance—but not always recognized as such—within my own worship tradition of Christianity.

How in particular might some traditional teachings and stories from this faith be freshly understood to address our present-day alienation from the land and mounting threats to this planet's community of life? What especially, in the face of such perils, might the "gospel," or Good News, of this faith tradition have to offer the world? These are questions that for some years I have been struggling to answer first for myself. And by "gospel" I mean not only the first-century *kerygma*, or core proclamation circulated by followers of the Christian "Way." For me, "gospel" pertains as well to how the Good News of salvation has been expounded throughout the Hebrew Bible and New Testament. I have in mind, too, its still more expansive character as interpreted, adapted, and applied to the changing circumstances of life throughout centuries of church tradition.

In reflecting here on what it might mean to embrace a "Green Gospel" suited to our time, I hope such reflection might therefore serve the further "eco-catechetical" purpose of encouraging readers, whatever their existing religious beliefs or training, to understand more deeply the theological foundations of Christian faith.

Such an inquiry into the nature of our relation to Creator, Creation, and all manner of ultimate questions can also benefit greatly, I believe, from an appeal to latter-day scientific findings and, as I'm typically drawn to highlight, diverse forms of artistic imagination. References to literary, liturgical, and scriptural texts—but also to select musical and visual creations—all play a role in the treatment that follows. Creative artistry can, I believe, not only convey but also help to reveal essential truths.

In that regard I am grateful for the original, wonderfully evocative artwork that Mary Priestley has contributed to this project by way of illustrations throughout the text. I am much indebted as well to Robin Gottfried and other fellowship leaders or participants in the Center for Deep Green Faith (https://deepgreenfaith.org/), kindred spirits who have for years stimulated my thinking about matters bearing on ecotheology and ecospirituality. Robin deserves special recognition for his founding and long-term guidance of the Center, as well as for his having conceived and overseen its unique certificate program in Contemplation and Care for Creation. Others I wish to thank for having contributed fruitfully, in one way or another, to my completion of this project include Arnold Benz, Christopher Bryan, Jerry Cappel, Elizabeth Carrillo, Collin Cornell, Cynthia Crysdale, Mary Foster, Julia Gatta, Phil Hooper, and Joyce Wilding. I owe much to Justin Hoffman, editor at Church Publishing Incorporated, for his early support of my hopes for the book. And I am especially grateful to editor Eve Strillacci at Church Publishing for her longstanding patience, encouragement, astute literary sense, invaluable suggestions, and total engagement in the process of bringing this volume to press. Whatever failings remain in the book, the final result, thanks to her, is surely something better than it would otherwise have been.

Introduction

Life cannot have been easy for Thomas Traherne, a lesser-known Anglican clergyman who served an obscure rural parish in the rolling hills of Herefordshire during the seventeenth century. The son of a shoemaker, Traherne was orphaned at an early age, grew up amid the turmoil of England's Civil Wars, and never married. Much of his writing, including some remarkable poems and prose meditations, remained unpublished or virtually unknown until our own day.

Yet the zeal with which Traherne acclaimed the beauty, glory, and wondrous giftedness of God's creation strikes me as unsurpassed. "We need nothing but open eyes," he wrote, "to be ravished like the Cherubims," to see the world as a divine temple and "the visible porch or gate of eternity." In his *Centuries of Meditations*, he insists that we are called to love and enjoy the earth, an abode suffused with God's goodness. "You never enjoy the world aright," he declares, "till the Sea itself floweth in your veins, till you are clothed with the heavens, and crowned with the stars." For "this visible world" we inhabit, seen through eyes of faith, resembles nothing less than "the body of God." As such, it is "wonderfully to be delighted in"—a "pomegranate indeed, which God hath put into man's heart."[1]

For me, Traherne's transformative vision remains an inspiration despite, or perhaps because of, the dispiriting degradation of the planet now taking place. The health of our common home, once imaged by astronomer Carl Sagan as a lovely but pale blue dot floating in the immensity of space, today looks

---

1. Thomas Traherne, *Centuries* (Wilton, Connecticut; and Oxford, England: Morehouse-Clarendon, 1985), 5, 18, 14, 65, 103.

more doubtful than ever since the dawn of human reckoning. As Sagan explains, his sense of the planet as merely a "point of pale light," a "lonely speck in the great enveloping cosmic dark," derived from an actual photograph of earth taken in 1990 by a far-flung *Voyager* spacecraft, billions of miles away from beyond our Solar System.[2] That photo presents us with a humbling spectacle, all the more so in light of the violence, waste, and accelerated spoliation of an extraordinary biosphere we know from closer view to be taking place there.

Many now find it hard, therefore, to sustain Traherne's assurance, consonant with Saint Peter's declaration while on the high mountain with Jesus, that it is "good for us to be here" (Mark 9:5), especially at this cultural moment when Christian faith, too, seems to have become an endangered species. Within the churches themselves, moreover, worshippers have rarely found occasion to appreciate precisely *how and why* an environmental vision should be understood as a central—not merely a supplementary or special interest—component of Christian faith and practice.

Yet the time may be ripe to recover such an appreciation. For while we can expect many of our neighbors would shrink from calling themselves "environmentalists," many more feel personally sustained by living amid green growing things, by glimpsing the presence of wildlife, and by contemplating scenes of natural beauty. Even those who are expressly committed to "earth care," however, and feel much anxiety about climate change and other forms of environmental degradation, may not connect their environmental concerns with religious faith. They may feel *moral* pressures toward the cause of environmental reform, especially in the face of climate change, but little sense of how—beyond the broad, rather shopworn motto of "stewardship"—such ethical claims could derive from

---

2. Carl Sagan, *Pale Blue Dot: A Vision of the Human Future in Space* (New York: Random House, 1994), 9.

Christian theology and spirituality. For what does earth care have to do with core elements of the gospel proclamation? With wonder, love, and praise? How might we understand an avowal of deep green faith to be not only relevant to but thoroughly rooted in gospel teachings, including those associated with traditional terms such as salvation, incarnation, God's reign, Transfiguration, the cross, and resurrection?

I believe these are live questions for many souls, particularly among active or prospective members of present-day faith communities. This book is meant for them. I have aimed, in ordering the treatment that follows, to address the relevant issues through a comprehensive logic that proceeds from first principles, through a sequence of chapters addressing key theological points, all the way to more applied reflection in the final two chapters. The word "theology" can sound intimidating or inaccessible at first, especially for those with little prior exposure to life within a community of faith. But in my first four chapters I have not hesitated to delve into those theological issues I find most germane to an environmentally informed and genuinely expansive worldview. For theology, like science, strikes me as providing an essential vision and matrix for understanding the complications of our experience in the wider world—a coherent modelling, if not an explanation, for how things really are beyond our immediate impressions, and for how they fit together.

Moving from ecotheology toward a more applied frame of reference in the fifth chapter, I consider there what it might mean to live out a Christian faith richly attuned to ecospirituality, with attention especially to contemplative practice and the formation of a personal rule of life. Finally, in the sixth and closing chapter, I probe the question of how deep awareness of the "green" features that already color the gospel traditions we have inherited in the Church might enrich our experience of liturgical worship, as epitomized by the eucharistic sacrament.

With the benefit of my own life-experience as a student and teacher of literature, and because many readers find it easier to apprehend otherwise abstract principles of theology or spirituality through portals supplied by artistic imagination, I have been moved to introduce at each stage, by way of illustration, a few telling literary excerpts. For many of us, artistic expression of all sorts offers an indispensable "way" toward encountering the deepest mysteries of life and faith. The music of Gustav Mahler, a favorite composer of mine, often strikes me as revelatory in this regard. "A symphony," Mahler once remarked, "must be like the world: it must encompass everything."[3] Mahler's *Symphony No. 3*, for example, dramatizes a universe story that reaches all the way from cosmic origins to move in its finale toward the sublime eternality of divine love. In figurative outline, at least, it encompasses everything. Sounded within the work's six movements is every register of creaturely and divine being, together with the whole course of evolutionary development.

The same must be true of any Christian faith worthy of the name. Faith must encompass everything—all things seen and unseen, human and nonhuman beings of every stripe, throughout the whole of creation. For us living today, a Jesus capable of rescuing just ourselves, or our kind alone, from sin and death can no longer be recognized as God's savior of the world. Only a cosmic Christ, as Saint Paul first envisioned, could possibly fulfill that role.

The "green gospel" of all-embracing salvation preached and accomplished by such an Anointed One surely qualifies as Good News. And that sense of a freshly liberating, hopeful, and existentially momentous proclamation conforms, of

---

3. Peter Franklin, *The Life of Mahler* (Cambridge and New York: Cambridge University Press, 1997), 173. Mahler reportedly made this statement, during a meeting with fellow composer Jean Sibelius, in response to Sibelius's seemingly more restrictive emphasis on the tight logic of symphonic form.

course, with biblical usage of the term "gospel" (*evangelion*). This book title's reference to a "Green Gospel" also has a contemporary coloring, insofar as the "green" qualifier means to situate present-day forms of environmental consciousness squarely within the New Testament's core teachings. Like St. Augustine's tribute to a divine beauty at once "so ancient and so new," the notion of a "Green Gospel" therefore strikes me as at once current in its idiom and ancient and elemental in essence. To call the setting of our own community of life a "pale blue" earth has paradoxical overtones as well. It suggests, on the one hand, the lovely hue of an oceanic world hospitable to life. Viewed from outer space roughly as far away as the moon, with scenes of terrestrial pain and conflict removed from sight, this blue planet wrapped in a stable, oxygen-rich atmosphere looks like a comforting home place indeed. And that is what we know it to be—an indispensable truth, albeit a partial truth. The famous Earthrise photo, from the Apollo 8 moon mission, allows us to absorb more fully this glorious view of planet earth.

On the other hand, we know Earth to be an exceedingly vulnerable body. It looks to be only one insignificant blue dot, after all, within an inconceivably vast blackness: a seemingly endless reach of lifeless space. Yet countless other bodies surround it; our planet continues to face bombardment from other celestial bodies, an exposure that has already triggered catastrophic extinctions and transformations in the past. We know, too, that what becomes visible closer to the ground is the severely compromised state of Earth's geophysical health—a point once underscored by poet Denise Levertov in lines titled "It Should Be Visible." Today it has become all too evident that, largely by human agency, climatic and other forms of environmental deformation threaten the flourishing and diversity of earth's biotic communities as never before since the dawn of human civilization.

In the face of all this, our species alone finds itself in a position to respond consciously to the "green gospel"—and, for that matter, should feel compelled to do so, as should our churches. Moreover, this evangelical imperative, though directed *toward* the likes of us, deserves to be heeded *for* the sake of all creatures inhabiting this pale blue planet.

My decades of teaching experience, during which I have endeavored to engage young people in matters of faith inquiry, as well as in reading and interpreting eloquent texts of environmental literature, convince me that many of these students find deep inspiration in contemplating their elemental and "original relation," as Emerson once termed it, to the green world that envelops them. Regardless of their prior religious training or lack thereof, and even amid our culture's rising tide of secularism, the robust current of spirituality flowing from texts by the likes of Henry Thoreau, Rachel Carson, John Muir, Willa Cather, Annie Dillard, Barry Lopez, and Wendell Berry clearly speaks to them.

True, the opening toward transcendence that contemplation of nature provides need not lead souls to gospel Christianity. But for an appreciable share of people today it can be a portal as promising as any other into lifelong engagement with a faith community. Pope Francis, drawing from Christianity's ancient fund of wisdom as well as present-day science, is one of those tracing a pathway there.[4] In any case, many of these newer citizens of the world will not be satisfied existentially with anything less than a vision of life which, like Mahler's symphonies or the Green Gospel, "encompasses everything." Neither, it seems to me, should the rest of us.

---

4. Pope Francis, with reference especially to the universal address of *Laudato Si': On Care for Our Common Home, Encyclical Letter* (Washington, DC: United States Conference of Catholic Bishops, 2015).

First Things

An All-Loving Creator and Our Place in Creation

Glimpses of Godliness in Creation

It is hard these days for any thoughtful person to avoid feeling despondent about the perils posed by climate change and other forms of environmental degradation to all forms of life on earth. But amid these woes, I sometimes find solace in bringing to mind a few memorable testimonials, drawn from my long engagement with the American literary tradition, to the surpassing marvel and grandeur of God's creation—on earth and elsewhere. *Creation* . . . that word evocative of where everything except God begins, and continues to be and to become, seems the right note on which to begin this inquiry.

So even in darker moods I find it heartening to recall the assurance of John Muir, gazing on the earthflow he perceived in Alaska's wild landscapes, that "this is still the morning of creation."[1] I am moved by the awestruck response of poet Denise Levertov to the great

> mystery
> that there is anything, anything at all,

---

1. John Muir, "Alaska," in *John Muir: Nature Writings*, ed. William Cronon (New York: Library of America, 1997), 679.

let alone cosmos, joy, memory, everything,
rather than void: and that, O Lord,
Creator, Hallowed One, You still,
hour by hour sustain it.[2]

And I take heart in remembering how fiction-writer John Cheever, even as he approached the end of a troubled, morally compromised life, could voice gratitude on behalf of us all for having known "that most powerful sense of our being alive on the planet," and of "how singular, in the vastness of creation, is our opportunity."[3]

But what, from the standpoint of Christian faith, does it really mean to regard "nature" as part of "creation"[4] rather than simply as everything that happens to exist—a universe based solely on natural laws, for reasons unknown and with no temporal point of origin? That everything in the ongoing flux of existence simply *is* could be considered a rough summation, in fact, of how Lucretius, first-century Roman philosopher and poet, had portrayed the universe in his *De Rerum Natura*. That the cosmos in some form had *always* existed corresponds, for that matter, with latter-day views of an eternal universe favored by some astronomers prior to emergence of the "Big Bang" theory promulgated by Georges Lemaître, James Hubble, and others in the 1920s.

To press this question of ultimacy still further, we might ask ourselves: what might it mean to believe that everything in nature is, and originally came to be, through the agency of an intelligent Being? This question arguably holds more

2. Denise Levertov, "Primary Wonder," in *Denise Levertov: Selected Poems*, ed. Paul A. Lacey (New York: New Directions, 2002), 192.

3. John Cheever, *Oh What a Paradise It Seems* (New York: Random House, 1982), 99–100.

4. For a particularly revealing treatment of this key question, with worthy emphasis on imagination's role in enabling us to apprehend nature as creation, see Norman Wirzba's *From Nature to Creation: A Christian Vision for Understanding and Loving Our World* (Grand Rapids, Michigan: Baker Academic, 2015).

existential significance for us than that of whether the material universe has always existed or has undergone a process of creation—however that process might have been impelled.

One need not be a Christian, or even a theist, to conceive of "Nature" as some sort of "Creation." All that's strictly required for this affirmation is a belief that the cosmos we inhabit *came to be*, in some manner or other. Perhaps even, one might imagine, through self-generative, natural processes yet to be explained scientifically. Or through an accidental confluence of generative circumstances, compounded then through the immense 13.8-billion-year span of evolutionary history. And while the prevailing "Big Bang" model of scientific cosmology is agreeably compatible with a theology of creation, it cannot be taken as "proof" of God's existence.

Empirical evidence cannot, in itself, assure us even of the existence of an unmoved mover, or faceless first principle, to say nothing of a loving and solicitous Deity. So what reasons of the heart might lead us to discern the presence of a divine Creator in the natural world? When might we be graced, in Wordsworth's language, to "see into the life of things"?[5] And what might such participatory perceptions,[6] combined with understanding drawn from Christianity's reasoned faith traditions, tell us about the identity of this Creator-God and the distinctive character of God's Creation?

In grappling with these questions, I am drawn to the resonant declaration of Gerard Manley Hopkins, whose poem, "God's Grandeur," announces: "The world is charged with the grandeur of God."[7] I find this bold statement confirmed,

---

5. Wordsworth, *William Wordsworth: Selected Poems and Prefaces*, ed. Jack Stillinger (Boston: Houghton Mifflin1965), 109.

6. For a fuller account of such perceptions, see Arnold Benz, *Astrophysics and Creation: Perceiving the Universe through Science and Perception* (NP: Crossroad Publishing, 2016).

7. Gerard Manley Hopkins, "God's Grandeur," in *Poems and Prose of Gerard Manley Hopkins*, ed. W. H. Gardner (Baltimore: Penguin, 1968), 27.

through centuries and across cultures, by a cloud of witnesses. For Hopkins, along with countless others of us, such a conclusion derives from experiential perceptions rather than purely rational evidence. These perceptions offer at least hints or glimpses, if not outward proof, that the created order enveloping us is really enchanted, despite rationalistic doubts to the contrary. Many of these intimations of a dimension beyond ordinary space and time even inspire a faith—surely consistent with scriptural revelation—that the material world we inhabit is animated by a Presence beyond ourselves.

Most of us, however, find ourselves sensing such a Presence only in certain privileged and often unguarded moments. These episodes of peak experience when we feel graced to "see into the life of things" are apt to occur at pivotal moments in our lives, including times marked by birth, death, or joyous reunion with loved ones. For many of us, such episodes can also be provoked, unexpectedly, through encounters with the nonhuman world.

What I recall about my own early glimpses of godliness in the natural realm reach back to childhood, extending into my adolescent years. During this period, my impressions of a numinous force in nature were particularly shaped by occasions when I felt blessed to breathe in the mountain grandeur of upstate New York's Adirondack Lake country during summers spent in that region. Its waters and high peaks remain today for me a land of spiritual refuge. I remember, too, how as a young adult I felt the godliness of creation coming into view as never before during a trip my wife and I took into the stark terrain of Joshua Tree National Park in Southern California. Standing outdoors there, late at night, I was startled to gaze upon the penetrating clarity and splendor of stars, planets, and constellations spread across the desert sky. A number of the astronomical bodies visible then—the Dippers, Polaris, Draco,

Cassiopeia—were already familiar to me. But to see them amid the silence and otherwise total darkness of this fierce, desolate land, displayed on a celestial dome so immense as to seem infinite in scope, was quite another matter.

How, I thought, could I ever have envisioned such a scene of transcendent beauty, brilliance, and magnificence had it not appeared to me in real life, before my very eyes? I realized in that moment that whatever belief system I embraced about my relation to the world would have to be expansive enough to take account not only of my own destiny and surroundings, but of this immeasurably grander vision as well. That has remained my conviction ever since. It has, if anything, been solidified by what I've learned about those countless worlds beyond our own from years of conversation with my friend Arnold Benz, an accomplished astrophysicist. For again, as Gustav Mahler once said of music's symphonic form, such a faith "must encompass everything."[8] We can perceive the creative flux of existence itself, in fact, as a kind of symphony—reflected, for example, in the song of whales, the music of the spheres, or the sounds we've lately discovered that trees share among themselves and with other living beings.

Glimmers of godliness in creation appear not only in scenes of grandeur, then, but also in the marvels that smaller creatures can reveal to us. In the light of scientific findings, I find it intriguing to realize, for example, how the invariably hexagonal pattern of a snowflake reflects, in all its beauty, the invisible ordering of its oxygen atoms and water molecules. The mating and mimicking habits of female fireflies are another such marvel, as is the amalgam of fungal life with alga or bacterium in vibrantly colored lichens. Each time I witness the familiar V signature of geese soaring overhead in the fall,

---

8. See Introduction, note 3.

I'm amazed to sense how the stamina and directional expertise of migratory birds surpasses any capacity of this sort we humans can muster, apart from our technology.

I perceive glimmers of godliness, too, in the way forest trees establish networks of subterranean connection with fungal communities and with other arboreal species. It seems that plants can indeed talk with each other, in a manner previously unsuspected except in a few traditional cultures.[9] One can also find hints, in the playful or masquerading behavior of some animals, that there is more depth to life on earth than any mechanistic, spiritless philosophy of nature could allow. Close to home for me, the provocative antics of my housecat beautifully illustrate that point. In sum, occasions abound for us to gain insight into the godliness of all creation.

It seems that even humans living in cultural settings dominated by the built environment may simply happen upon such insights without deliberately seeking or expecting them. American sage Ralph Waldo Emerson, for example, writes of having been seized by a Transcendental sense of joyous union with Nature while "crossing a bare common, in snow puddles, in twilight, under a clouded sky, without having in my thoughts any occurrence of special good fortune." The exhilaration Emerson felt then blended strangely with apprehensions of the sublime, leaving him "glad to the brink of fear."[10] Yet those engaged in some form of spiritual discipline and practice are evidently predisposed, at least, to "see into the life of things," as are those open to an expansive exercise of imagination.

---

9. My Sewanee colleague David George Haskell has fruitfully explored these matters in books of his, including *The Forest Unseen: A Year's Watch in Nature* (New York: Viking, 2012) and *The Songs of Trees: Stories from Nature's Great Connectors* (New York: Viking Random House, 2017).

10. Ralph Waldo Emerson, *Nature*, in *Ralph Waldo Emerson: Essays & Lectures*, ed. Joel Porte (New York: Library of America, 1983), 10.

At base, the human faculty of imagination amounts to much more than just an entertaining mode of make-believe. As memorably described by Samuel Taylor Coleridge, imagination is that vital faculty of mind by which we envision the wholeness of reality, the often-unseen web of connections that unifies otherwise disparate elements of God's creation. It is an impulse we might regard as inherently ecological. It is a re-creative, "living Power and prime Agent of all human perception" that echoes, as Coleridge reminds us, "the eternal act of creation in the infinite I AM."[11] Imagination must therefore play a role in those moments of transparency when we are privileged to look iconically not only *upon* but *through* features of our material world to a transcendent or "numinous" reality. If humanity as a whole ever sees a pathway that moves past its current complacency in the face of climate change, such a breakthrough might come about only through a collective awakening of imagination.

Can we, however, understand the Presence discerned in these epiphanic moments to be personal and loving, albeit in a manner surpassing anthropomorphic projections? There are admittedly vast differences among these three prospects: a) belief in an abstract Presence undergirding existence; b) belief in a personal God; and c) belief in the kind of God revealed through the teachings of biblical Judaism and Christianity. Is there something like a face, a discrete identity, assignable to the Creator of all things? And if we are to believe that "God is love" (1 John 4:16), does this Creator truly *love the world*—something an impersonal first principle of cosmic generation could scarcely be expected to do?

It seems that only a personal God, if One exists, could sustain any kind of affective, personal relation to other beings. For Gerard Manley Hopkins, in his closing stanza of "God's

11. S. T. Coleridge, *Biographia Literaria*, vol. 1, ed. J. Shawcross (Oxford: Oxford University Press, 1973), 201–202.

Grandeur," the presence animating all life is plainly that of divine Love. Despite the persistent failure of our species to honor the godliness of God's earth, Hopkins does not hesitate to celebrate the Creator's maternal warmth and fond solicitude:

> And for all this, nature is never spent;
> There lives the dearest freshness deep down things;
> And though the light off the black West went
> Oh, morning, at the brown brink, eastward springs—
> Because the Holy Ghost over the bent
> World broods with warm breast and with ah!
> bright wings.

In the cosmological vision of Dante Alighieri, this same Love "moves the sun and the other stars."[12] Wisdom teachers from both East and West have seen this Love written across the whole face of creation. Naturalist John Muir believed that "God's glory is over all his works, written upon every field and sky." But after spending his first winter in California's Yosemite Valley, he rather colorfully insisted that "here it is in larger letters—magnificent capitals."[13] For him, then and there, the divine Presence was unmistakable.

For followers of the Way, according to New Testament usage, a personal face of such Presence can best be discerned in the Christ of God. For example, in visions of the medieval mystic Julian of Norwich, Christ's "homely loving" enfolds "all that is made." The amplitude of this caring encompasses, in other words, the whole of creation, rendering it as comparatively small as a hazelnut—or, we might say with the benefit of modern science, as the minuscule size of the entire

---

12. *Dante's Paradiso*, ed. and trans. John D. Sinclair (New York: Oxford University Press, 1961), 485.

13. John Muir, in *John of the Mountains: The Unpublished Journals of John Muir*, ed. Linnie Marsh Wolfe. (Boston: Houghton Mifflin, 1938), 47.

physical cosmos prior to its colossal expansion from the Big Bang. For Mother Julian, the essential meaning of everything might therefore be figured in a definitive monosyllable. As she intones in her *Revelations*, "Love was his meaning. Who showed it thee? Love. What showed he thee? Love. Wherefore shewed it thee? For Love."[14]

The Creative and Continuous Hue of Creation

Is it then possible to prove, scientifically and definitively, the existence and character of a Creator-God? Many in the Western world, invoking the presumed authority of "natural religion," once believed so. But in the light of our most estimable present-day knowledge, I think not. It is at any rate vain to suppose that our notion of divine creation can be plausibly reduced to a so-called "God of the gaps"—that is, to whatever cosmic enigmas science finally seems unable to explain.

Even or especially in our own day, we may indeed be wonderstruck by certain marvels in the ordering of cosmological existence, including what is known in scientific circles as the "fine-tuning" or "anthropic principle."[15] That the universe we inhabit happened to be one in which life could flourish on at least this planet if not others was, in other words, far from inevitable. The principle of "fine-tuning" suggests that it's not only the peculiar, statistically improbable character and orbital location of earth itself that defies expectation. The principle applies as well to the comparatively rare makeup of our solar system, of the cosmos as a whole, and a long list of constants required to be operative by the laws of physics for life and a habitable planet to exist. Had the strength of gravitational

---

14. Julian of Norwich, *Revelations of Divine Love*, adapted by Dom Roger Hudleston, OSB (Mineola, New York: Dover, 2006), 8–9, 169.

15. Further explained by Arnold Benz in *Astrophysics and Creation*, xii, 152, 167–170.

force been other than it is, for example, or if the early universe's rate of galactic expansion were either greater or less than it happened to be, or if the mass of electrons were other than it is, none of us would be here today to ponder such things.

If we reflect on these marvels with eyes of faith, we might even see in them signs or "traces" of the God known through Christian revelation. Our participatory perception of them offers fresh insight, if not proof.

Still, I find it quite sensible to conclude from a melding of evidential reason, faith, and experience that the universe we inhabit is not only ordered but meaningful. Not only meaningful but subsisting in and through the agency of an eternally creative Being. Does it make sense, then, to describe this faith-inspired vision of divine creation as "intelligent design"? Or if not, why not?

A major shortcoming of the "intelligent design" theory of cosmogenesis advanced by some in the United States is the faulty supposition that such an approach can supply, on scientific grounds, an effective refutation of evolutionary science. While proponents of intelligent design tend to be earnest believers who mean to defend the faith without direct appeal to Christian revelation, their terminology does not, ironically, align well with most biblical portrayals of the Supreme Being. As John Polkinghorne and Nicholas Beale observe:

> Of course God is "intelligent," and of course teaching in schools that "godless evolution" is the only story is intellectually and spiritually impoverishing. But evolution is no more "godless" than gravity or electromagnetism.
>
> God is never spoken of as a "designer" in the Bible: he is Creator and Father, and a Father does not "design" his children. Even a great creative writer does not exactly "design" her or his characters . . . By

endowing us with free will and giving us the capacity to love, God calls us to be in a limited but very important sense co-creators.[16]

Far from bolstering esteem for the Creator-God, the ideology of intelligent design fails to recognize the richly creative, cooperative, and dynamic force inherent in the Creator's inspiration of evolutionary processes. Even if we affirm, in faith, that God is the first cause and ground of all creation, we must acknowledge—as intelligent design does not—the substantial role and freedom that God has allowed within the cosmic drama for all manner of secondary causes and chance developments.[17] Such causes, though sometimes agents of sorrow and malignancy, also infuse into existence a welcome color, variety, beauty, and unanticipated marvels.

In sum, the Creator-God of intelligent design is not genuinely creative—not, at least, by analogy with the sublime expressions of creativity we have come to recognize in great literary authors, composers, public leaders, and painters. Nor is the Designer-God artistic, vital, original, or playful. What this God designs, in figurative essence, is just a series of static blueprints, destined for sequential realization in the material realm. These preconceived blueprints for the design of all creaturely existence might be correct in every detail but comparatively lifeless, loveless, and cheerless in their conception.

In contrast to such a predesigned, blueprint model of creation, we might consider how, in the Wisdom tradition of the Old Testament, an enigmatic female personification of

---

16. John Polkinghorne and Nicholas Beale, *Questions of Truth: Fifty-one Responses to Questions about God, Science, and Belief* (Louisville, Kentucky: Westminster John Knox, 2009), 57.

17. For a discerning discussion of how this matter of secondary causality relates in Christian perspective to God, contingency, chance, and probability, see Cynthia Crysdale and Neil Ormerod, *Creator God, Evolving World* (Minneapolis: Fortress Press, 2013).

Wisdom once sported with God in playful "delight, rejoicing before him always" in the elemental acts of creation, "rejoicing in his inhabited world and delighting in the human race" (Proverbs 8:30–31).

And in the Book of Job, the final voice from the whirlwind recalls how the divine artisan's genesis of all things had melded with a glorious music, "when the morning stars sang together and all the heavenly beings shouted for joy" (Job 38:7).

That note of musicality, of divine creation as a ceaseless and soul-stirring flow rather than a static design, has been echoed since by countless writers and lovers of nature. John Muir, for example, heard such music not only in the visible flow of California watercourses but also in what he perceived to be the geological origin of Yosemite's spectacular features through the movement of glaciers. "Contemplating the lace-like fabric of streams outspread over the mountains," he wrote, "we are reminded that everything is flowing—going somewhere, animals and so-called lifeless rocks as well as water."[18] The free flow of nature's constituent creatures and energies that is envisioned here reflects, moreover, that radical freedom with which God has likewise endowed human nature and all formative processes of evolutionary change.

Muir seems never to have lost the sense of living in a world and time when creation was still "just beginning, the morning stars 'still singing together and all the sons of God shouting for joy.'" The dynamism inherent in this vision of things propels the sort of breathless, exclamatory rhetoric often seen in Muir's writing, as when he declares that "From form to form, beauty to beauty, ever changing, never resting, all are speeding on with love's enthusiasm, singing with the stars the eternal song of creation."[19]

---

18. Muir, *My First Summer in the Sierra*, in *John Muir: Nature Writings*, 292.
19. Muir, *My First Summer*, 278, 226.

Creation, without benefit of human language, speaks to us by virtue of its own dynamic presence. In biblical terms, it testifies from the Book of Genesis onward to "the glory of God" and offers a form of speech that "goes out through all the earth" (Psalm 19:1–4). In *The Magician's Nephew*, the action of which anticipates the rest of C. S. Lewis's *Chronicles of Narnia*, the leonine voice of the character Aslan surpasses ordinary speech to become wild but beautiful song, as Aslan joins in harmony with other voices and "the voice of the earth itself"[20] to sing the land of Narnia into existence. This poetically stirring account again echoes the Book of Job, in portraying divine creation as that which reaches beyond words to animate the music of existence.

An especially cogent literary case for why creation should *not* be equated with God's having designed a fixed blueprint or clockwork for existence appears in Annie Dillard's nonfictional classic, *Pilgrim at Tinker Creek*. There, in a paragraph that muses upon how to reconcile our will to believe with all the troubling and apparently inhospitable features of the natural world that Dillard exposes elsewhere, she leads us toward the startling image of a Lord who "loves pizzazz":

> The point of the dragonfly's terrible lip, the giant water bug, birdsong, or the beautiful dazzle and flash of sunlighted minnows, is not that it all fits together like clockwork—for it doesn't, particularly, not even inside the goldfish bowl—but that it all flows so freely wild, like the creek, that it all surges in such a free, fringed tangle. Freedom is the world's warp and weather, the world's nourishment freely given, its soil and sap: and the creator loves pizzazz.[21]

---

20. C. S. Lewis, *The Magician's Nephew* (New York: HarperCollins, 1983), 106.

21. Dillard, *Pilgrim at Tinker Creek* (New York: HarperCollins, 1999), 139.

I think it telling that present-tense verbs prevail in this account. Dillard reports that God "loves pizzazz," overseeing a creative process, still in play, wherein "all flows so freely wild." Creation should not be misconstrued as purely a one-time event. Star formation continues to take place, as does biological evolution on many fronts. New stages of geological shaping and of human history remain in play. From the standpoint of faith, too, the continuous rather than strictly generative character of God's creation is a principle worth taking to heart. And like the principle of God's creation *ex nihilo* (that is, from nothing), it is an idea with ample foundation in Scripture and Church tradition.

Within the round of daily life, we find it easy enough to recognize that new things and lives are constantly replacing older ones, whether we wish it so or not. Continuous creation can scarcely be ignored on this scale. How can it be, I sometimes wonder, that my granddaughter, who had not even been born nine years ago, is now entering swim races, besting me in chess games, and reading aloud some of my favorite stories? How can I ignore each day the never-ending signs around me of new birth and new death, human or otherwise? Yet when bringing to mind certain monumental features of the natural world, including celestial bodies, we are more apt to harbor illusions of permanence. It is probably harder still to grasp what the ceaseless, divine flow of creation might look like envisioned on a global or even cosmic scale.

The nearest I can come to imagining such a thing is represented in *Walden's* penultimate chapter, "Spring," in which Henry David Thoreau shares an account of what he observed in a thawing sandbank constructed for the Fitchburg railroad line in Massachusetts. Surprisingly, the author's observations here center not so directly on conventional signs of the season in the green world all around him at Walden Pond but on what amounted to a common artifact of industrial civilization in nineteenth-century America. The passage in question describes the flow of sand and clay during spring thaw on the bankside of a railroad cut, at the western edge of the famous pond in Concord, Massachusetts, beside which Thoreau had built and inhabited his own dwelling.

What Thoreau envisions in this common bank of sand, however, is nothing less than the whole course of earth's evolutionary history, sustained into the present moment with creation still in motion. He first observes how the icy sand, warmed by sunlight, flows down the slope "like lava." The cosmogonic story continues as the sandstreams seem to form themselves into leaves and vines. Then, swept by the stream of multilingual correspondences emanating from the word "lobe," these vegetative leaves turn into fatty leaves suggestive of animal parts and, finally, of the human body. For after all, he reflects, what "is man but a mass of thawing clay?"[22] Before Thoreau's eyes, the world develops almost instantly—as if surveyed through time-lapse cinema—from chaos to cosmos, from primordial energy and molten matter to the leaves of his own book-in-progress. He witnesses Creation progressing from lava sand, through vegetable leaf, all the way to human consciousness, imaged in the great tree of language. Within the account as a whole, one can trace this development through an alliterative sequence of "L" words: from lava through leaf, lobe, and language.

The visionary and linguistic tour de force represented in this encapsulation of earth's life-story is at once colored by Thoreau's extensive reading in the "new science" of his time[23] and infused with his sense of the sacred. Although Thoreau was not a professed Christian, his portrayal of a Creator-God who is sportive, resourceful, and extravagantly creative is

---

22. Henry David Thoreau, *Walden*, ed. Stephen Fender (New York and Oxford: Oxford University Press, 1999), 271–275.

23. I discuss this noteworthy backdrop to the visionary outlook Thoreau brings to the sandbank passage in *Making Nature Sacred: Literature, Religion, and Environment in America from the Puritans to the Present* (New York: Oxford University Press, 2004), 133–142.

compelling. It also recalls the vibrant force of Lady Wisdom, as set forth in the Hebrew Bible.

Most striking, perhaps, is the conviction Thoreau conveys that he has become a first-hand witness to the divine creation taking place not only once upon a time, but unmistakably in the here and now. He testifies to feeling as though he stood indeed "in the laboratory of the Artist who made the world and me—had come to where he was still at work, sporting on this bank, and with excess of energy strewing his fresh designs about." He finds "this one hillside," illustrating "the principle of all the operations of Nature." He sees the entire life story of our biosphere unfolding from an elemental pattern we might today liken to DNA—that is, from how "the Maker of this earth" simply "patented a leaf." But who "will decipher this hieroglyphic for us," Thoreau goes on to ask, "that we may turn over a new leaf at last?"[24]

Since Thoreau was no champion of the rail commerce that had begun in his day to transform American landscapes, it is all the more surprising that he chose to dramatize the unfolding of all Creation on Earth by focusing attention here on an ordinary railroad bank. Only a writer of rare talent would think to map onto this commercial construct of sand and clay his panoramic survey of geological, biological, and cosmological evolution. In doing so, Thoreau also identifies the ongoing emergence of Cosmos from Chaos and the "Maker of this earth" with a joyously extravagant Creator.

Strangely, too, the author suggests that what he finds displayed on the sandbank amounts to a "hieroglyphic." And as he surely realized, the word's literal sense, from its Greek etymology, is "sacred writing."

But to discern the meaning of this sacred script so copiously spread across the face of nature requires reflection, imagination, and an experiential familiarity with the green world that

---

24. *Walden*, 273–275, 279.

many in our culture of rootless civilization seem to lack. Who can help us decipher the signifying pattern and inner life of a leaf? Who, for that matter, can guide us toward comprehending the language and spirit operative within every other aspect of the natural world? "What Champollion," Thoreau had asked, can decipher the unspoken language of Nature's own hieroglyphic? For just as the hieroglyphic language of the Rosetta Stone remained impenetrable to moderns until Jean François Champollion, the French Egyptologist, managed to break its linguistic code, so also we need to rely on others, including present-day scientists and naturalists like Thoreau himself, to help us interpret the hiero-glyph that is always and everywhere being inscribed within material reality. Church history confirms that spirit-filled exegetes can often help us see more deeply into the truths of scriptural revelation. By the same token, I believe we owe much to those souls, both living and dead, gifted to interpret for others the wealth of sacred script inherent in the created order, a record esteemed throughout much of Christian tradition as God's Second Book of Revelation.

Why a Triune God?

Much of the appeal of living, working, or playing out of doors is the chance to recover our elemental linkage to soil. As gardeners or farmers, forest hikers or birders, and many others of us have long realized, a richly sensate satisfaction can be found in realizing our kinship with the manifold creatures and material features we share with the larger community of life on earth. In the face of such earthy experiences, we may not be inclined to suppose that our understanding of creation could possibly be enhanced through the seemingly abstract medium of theological teaching. Among all core teachings of the Christian faith, none is apt to be regarded as more abstract and conceptually elusive than the doctrine of a Trinitarian

Deity. So why a Triune God? Even in Christian faith communities where Trinitarian formulas are regularly expressed in liturgical usage, the existential import of this article of faith—within one's overall life of faith, to say nothing of one's ecospirituality—rarely seems to gain prominence.

To appreciate more fully the relevance of Trinitarian theology to our apprehension of a creation "charged with the grandeur of God,"[25] I believe we must first challenge two doubtful but familiar impressions of this doctrine. One is the supposition that the Trinity is a forbiddingly abstract and abstruse doctrine—hence one that might best be left unmentioned in public discourse, aside from the obligatory address on Trinity Sunday. Another misconception is to approach the Trinity as inherently problematic rather than illuminating. If regarded mainly as an esoteric puzzle beyond our capacity to solve, then this doctrine looks indeed to be a barrier rather than a breakthrough to fuller understanding.

But without reviewing the history that eventually led to Church teaching on the Holy Trinity as officially defined at fourth-century ecumenical councils, I think it fair to say that this doctrine did not arise from someone's armchair philosophizing. Part of its impetus was the pressing need for Christian worshippers of God to clarify the relation between Jesus, as the Christ of God, and the Father of whom Jesus spoke. But it involved as well a collective need, grounded not in abstraction but in life experience, to better comprehend the relation between Creator and Creation. While Trinitarian views had already been broached, without elaboration, in New Testament writings, a cohesive teaching could develop only over time through the faith community's joint experience and reflection on the matter.

What human beings, Christians or otherwise, experience at peak moments of encounter with the greater world around

---

25. Gerard Manley Hopkins, "God's Grandeur," 27.

them often generates a paradoxical sense of the Supreme Being as both immanent and transcendent—that is, as an almost palpable presence here and now and yet ineffable, eternal, a Self surpassing anything or anyone in creation. How is it, for example, that as I sit quietly to watch a cascading watercourse somewhere in Yosemite Valley or the Adirondack Mountains, I sense a flow of divine spirit that seems to be at once *within* and yet *beyond* the physical scene I am trying to absorb? The way we experience God in these moments cries out for something more than an either/or solution: God beyond us or God within us. Trinitarian belief thus offers a salutary frame of reference for absorbing some of the deepest paradoxes of existence. Like other revelations opening into sacred mystery, the Trinitarian doctrine provides no definitive solution to life's enigmas but should be regarded as more nearly an advance of understanding, a grammar for a language we can scarcely speak, than as a problem.

The familiar hymn, "St. Patrick's Breastplate," is an expression of Trinitarian spirituality noteworthy for its display of an earthy realism and ardor that belies the esoteric abstraction often attributed to this doctrine. The Gaelic poem in question, otherwise known as "the Deer's Cry," is thought to date from the late seventh or early eighth century. It is therefore unlikely to have been composed by Patrick himself. Yet it reflects a Celtic world in which God could readily be invoked as a presence coterminous with everyday life.[26] In this same world, also a vulnerable setting exposed to grave physical perils, a Spirit-filled Creation felt as close to hand as the windswept boulders and oceanic wilds that defined much of everyday existence on these island edges of civilization. In such a place, it required little effort to bring to mind—and take to heart—earth's elemental features of rock, wind, sun, and sea.

---

26. A cultural and spiritual atmosphere effectively described by Esther de Waal in *Every Earthly Blessing: Rediscovering the Celtic Tradition* (New York: Morehouse Publishing, 1999).

I bind unto myself . . . the whirling wind's tempestuous shock,
the stable earth, the deep salt sea
around the old eternal rocks.

Citation from some selected stanzas of the hymn, as translated by Cecil Frances Alexander, can begin to dramatize these
points:

> I bind unto myself today
> The strong Name of the Trinity,
> By invocation of the same,
> The Three in One, and One in Three.

> I bind this day to me for ever,
> By power of faith, Christ's incarnation;
> His baptism in the Jordan river;
> His death on cross for my salvation;
> His bursting from the spicèd tomb;
> His riding up the heavenly way;
> His coming at the day of doom;
> I bind unto myself today.

> I bind unto myself today
> The virtues of the starlit heaven,
> The glorious sun's life-giving ray
> The whiteness of the moon at even,
> The flashing of the lightning free,
> The whirling wind's tempestuous shocks,
> The stable earth, the deep salt sea,
> Around the old eternal rocks.

> Christ be with me, Christ within me,
> Christ behind me, Christ before me,
> Christ beside me, Christ to win me,
> Christ to comfort and restore me,
> Christ beneath me, Christ above me,
> Christ in quiet, Christ in danger,
> Christ in hearts of all that love me,
> Christ in mouth of friend and stranger.

I bind unto myself the Name,
The strong Name of the Trinity,
By invocation of the same,
The Three in One, and One in Three.
Of whom all nature hath creation,
Eternal Father, Spirit, Word:
Praise to the Lord of my salvation,
Salvation is of Christ the Lord.[27]

The hymn reflects an earthy form of meditative piety familiar in Celtic practice. In it, one finds the replicated, mantra-like invocations—verging on incantations—of Christ before, behind, beside, and within the speaker conjoined with unseen spirit creatures, as well as with a vast communion of human saints. Within the Trinitarian frame of this all-enfolding formula, speakers shield themselves both with the God of Christian revelation and a Spirit-filled Creation, vividly pictured in "the whirling wind's tempestuous shocks, the stable earth, the deep salt sea, around the old eternal rocks."[28]

From the standpoint of this hymn, the Holy Trinity is a "strong name," to be wielded with confidence for one's solace and protection in an often-hostile world. Of a culture still colored by Druidic paganism, the speaker in the hymn seems haunted at times by apprehensions of wizardry, poison, and unforeseen violence. Set against these fears, the Trinity's "strong name," equivalent to a formidable shield or Gaelic "lorica," is eminently practical. It is a name bound up with the force of "lightning free" and those "old eternal rocks." As such, the Three in One is conceived to be not just a speculative idea but a force to be reckoned with. Today's readers, too,

---

27. The language here, from Cecil Frances Alexander's rendering of the Gaelic, is that found in selection #370 of the Episcopal Church's *The Hymnal 1982* (New York: Church Hymnal Corporation, 1985).

28. Ibid.

might identify with the hymn's aspiration toward full, personal absorption of the trinitarian mystery, centered in One from "Whom all nature hath creation."[29]

As theologian Jürgen Moltmann has cogently explained, the doctrine of God as One-in-Three persons should thus be seen as pivotal to a faith-centered vision of ecology, as well as to Christian faith more generally. That is so because of the light it casts on the nature of God in relation to the whole of created nature. The vision of a Triune God contributes most critically toward shaping a robust ecotheology, I believe, by *holding together, in creative and paradoxical tension, two seemingly contrary notions of the Godhead.* Or as Moltmann puts it, the "trinitarian concept of creation binds together God's transcendence and immanence," thereby conjoining the partial, opposite truths represented both in radical monotheism and in a pantheism that would virtually *equate* Nature with divinity.[30]

Is the Creator-God to be envisioned, in other words, as substantially *in* creation or as a Being quite *above and apart* from it? The prevailing sense that Christianity shares with Judaism, Islam, and other religious traditions calls for us to recognize the partial truth of both perspectives. But how to draw the two together into an organic but paradoxically textured unity is a challenge which, in Christian terms, Trinitarian theology is ideally suited to address.

For Christians, a belief that the Creator-God is substantially *in* Creation finds confirmation not only in certain of our perceptions about the natural order, as already noted, but also in the doctrine of God's creaturely incarnation and

---

29. Ibid.

30. Jürgen Moltmann, *God in Creation: A New Theology of Creation and the Spirit of God*, trans. Margaret Kohl (New York: HarperCollins,1991), 98. Another useful theology, linking trinitarian thought not only with the heritage of Athanasius and Aquinas but also with a latter-day, evolutionary view of creation is that expounded by Denis Edwards in *Partaking of God: Trinity, Evolution, and Ecology* (Collegeville, Minnesota: Liturgical Press, 2014).

self-emptying in the person of Jesus. Yet to suppose God to have been so thoroughly subsumed into the natural order as to erase all distinction between the Creator and Creation—leaving us, in effect, with the unorthodox religious outlook of "pantheism"—is unsettling. For surely the nonhuman world, to say nothing of human nature, presents too many discontinuities, faults, and downright horrors to demand our worship as unqualifiedly divine. Who among us really wants to worship disease-bearing bacteria, or the devastating sweep of earthquakes and monsoons?

The opposite model of God's relation to everything else, which envisions a Creator radically *detached* from Creation, has still other drawbacks. It fails to accord with much of the tenor or content of scriptural revelation. Moreover, such a "monarchical monotheism,"[31] as theologian John Macquarrie once labelled it in contrast to a preferable "organic model," offers less encouragement to love nature and preserve the earth than pantheism does, or other forms of monotheism. There is much to recommend that more nuanced monotheism, sometimes called "panentheism," which credits some of the immanence claims of diverse polytheisms or nature religions while honoring the core principles of Abrahamic faith.

As Moltmann therefore concludes,

> The trinitarian concept of creation integrates the elements of truth in monotheism and pantheism. In the panentheistic view, God, having created the world, also dwells in it, and conversely the world which he has created exists in him. This is a concept which can really only be thought and described in trinitarian terms. . . . The de-divinization of the world has progressed so far that the prevailing view of nature is

---

31. John Macquarrie, "Creation and Environment," in *Ecology and Religion in History*, ed. David and Eileen Spring (New York: Harper and Row, 1974), 48–75.

totally godless, and the relationship of human beings to nature is a disastrous one. This means that today we have to find an integrating view of God and nature which will draw them both into the same vista. It is only this that can exert a liberating influence on nature and human beings alike.[32]

Three distinctive features of Trinitarian belief deserve mention, finally, as disposed toward enhancing faithful environmentalism. To begin with, the Trinitarian model is in essence relational and communitarian. It is also personal, at least in a figurative sense apart from our usual preconceptions, and inherently dynamic.

The Trinity's relational character becomes clear insofar as we are to understand that the persons traditionally named as Father, Son, and Holy Spirit constitute a vibrant community of love in themselves, even apart from all that God wills to create. In a colorfully fictive version of the universe story penned recently by Robert Gottfried, a singer, drummer, and bassist join forces for the sake of making beautiful music. This allegorized ensemble, the Home Jazz Trio, takes pleasure at first in sounding the harmony of existence just in and for themselves.[33] We are likewise to understand that it is only out of loving desire, with no hint of need or obligation, that the Holy Trinity willed to extend relationality through the ongoing formation and sustenance of a universe teeming with energy and all kinds of creatures.

It is difficult to speak of the Trinity's "Persons" while affirming their consubstantial unity, and without attributing to them human characteristics we know to be inapplicable. Still, as the titles Father and Son press us to acknowledge, some

---

32. Moltmann, 98.

33. Robert Gottfried, *The Audacious Great Singing School Adventure: A Fable* (St. Petersburg, Florida: Booklocker.com, 2021).

analogy with loving interchanges in the human realm must figure in our attempts to picture the self-contained or so-called "inherent" Trinity of Persons. Further complicating the matter, in addition to some gender questions, is the need to avoid a narrowly modalist attribution of functions to each Person of the Trinity. It is today customary, and perhaps even appropriate in broadest strokes, to speak of the Father as Creator, the Son as Redeemer, and the Spirit as Sustainer. Yet scriptural sources should discourage our attaching a discrete function too readily to each Person since, for example, in revealing passages from St. Paul and St. John, we find portrayed as Creator not only the Father, but also Christ the Son. At the same time, there are corresponding drawbacks to avoiding "person" language altogether in Trinitarian discourse. To speak of God as an "it" is to conceive of a Supreme Being far removed from the solicitous God of Abraham and his descendants. And of course no impersonal force, no "thing," can possibly be expected to love.

Neither should we picture the Trinity as a still-life portrait. The dynamic interplay of love shared among Father, Son, and Spirit is an appealing attribute of the Holy Trinity. For the Greek Church fathers especially, such activity reflected a kind of supernal harmony, even a joyous dance—or *perichoresis*—that is sustained eternally. It models, too, a generative energy that extends throughout the larger bounds of creation.

We, too, are invited to join in the dance of life—a joyous rhythm mirrored in the rush of subatomic particles or in the ongoing symphony of kinesis performed by cosmological bodies. The late Michael Mayne, in his eloquent reflection on matters contemplated toward the end of his life, says this about his own stumbling efforts to join the dance: "Despite the dark and testing times, the sickness, the failures and the disappointment, I shall respond with a heartfelt 'Yes!' if that Jewish rabbi is right and God's first question at the judgment will be 'Did you enjoy my creation?' for it has been a richly diverse and fulfilling life."[34] The dance imagined to persist in and by Persons of the Holy Trinity thus looks to be paradigmatic of much else that moves under the sun.

At one level, though, we have to regard this movement within the Godhead as literally inconceivable. For in the only world we know, movement or change of any kind can exist only in relation to some temporal standard. Within a timeless frame of reference, then, how might one envision the kinetic flow of *perichoresis*? We cannot hope to know, this side of the grave, yet the dynamism attributed to the Holy Trinity bears some fruitful implications for ecotheology.

Consider, for example, how the opening, Priestley version of the creation story in the Book of Genesis just might, read in isolation, suggest to readers that God, after six days of intense involvement in his newly formed cosmos, had retreated first into total rest, then into a stasis only occasionally interspersed thereafter with episodes of active intervention. On other grounds, too, one might once have supposed that while the creation is subject to perpetual alteration, God remains fixed in a static realm, unmoving in essence.

Yet we have better reason now to reach a contrary assessment: that there lies even within God a ceaselessly vital, interlayered, and active community of life and love—though not necessarily in the anthropomorphic mold often favored by process theologians. This vibrant view, effectively amplified by Trinitarian doctrine, matches what I take to be the overall sense of biblical teaching. The disposition of a dynamic God who dances, and Who "loves pizzazz," also meshes beautifully with a created order we know to be undergoing multiple forms of evolutionary transformation. That transformation remains in play though we are now presumed to be living, by some reckonings, in a post-Resurrectional eighth day of creation.

---

34. Michael Mayne, *Learning to Dance* (London: Darton, Longman and Todd, 2011), 230.

Our Place in Creation

Surprisingly, perhaps, these questions about the character and implications of belief in a Trinitarian God actually do, from the standpoint of orthodox Christianity, link up directly with the question of where we as mortal humans stand within God's created order. For me, that linkage is most memorably portrayed in literary terms by Dante, at a critical moment of the final Canto (XXXIII) in *Paradise*, from the *Divine Comedy*. It comes as Dante's pilgrim-narrator approaches the end of his long, often arduous and unsettling journey through hell and purgatory, toward the summit of Paradise. There, having been guided along the way first by Virgil, then Beatrice, and finally by St. Bernard of Clairvaux, he is privileged to receive at least a mediated glimpse of the Beatific Vision. That vision includes, in addition to the Celestial Rose peopled by the communion of all saints, God's great book of Eternal Light, binding with love within its leaves all that is or ever was of Creation.

The revelation culminates in a visual simulacrum of the Divine Trinity itself. Three different yet wholly interfused circles of colored light strike the pilgrim-narrator's eyes. But he is all the more amazed to find painted there, amid the supernal revelation, "our likeness" (*nostra effige*)—that is, the image of our own humanity.[35] Who would have thought to find an image of *us*, messy and erring creatures that we are, lodged within this portrait of the Supreme Being. It's a little like our discovering suddenly, decades later, that a very ordinary, unassuming childhood friend of ours from the same neighborhood had under another name risen to become the most powerful and respected person on earth. Who would have guessed?

Of course the doctrinal logic behind this apparent anomaly lies in the Person of Christ, understood to conjoin within

---

35. *Dante's Paradiso*, trans. John D. Sinclair, 484–485.

his dual nature true divinity together with that of true and full humanity. As a consequence, the whole of creaturely humanity is effectually elevated to glory as well—as ancient teaching, especially among theologians in the Eastern Church, dared to affirm. The doctrine of *theosis*, which recognizes the godliness embedded within human nature, derives in turn from Biblical assurances that we are made in God's image. Just as God emptied God's divine self to be *in* Christ, who shares our human nature, so also we have been made kindred with and *in* God. One Christian response to the question of our placement within the cosmic scheme of things would, therefore, consent see us seated on high, with and in the Second Person of the Divine Trinity.

Yet this joyous but partial truth must be qualified by the contrary truth that we, quite unlike the Creator, also happen to be blameworthy, arrogant, decidedly imperfect citizens of God's Creation. We are certainly not God, though too often inclined to behave either individually or collectively as though we were. As animal primates not far removed in geological time from our emergence on the savannahs of Africa, we not only *belong* to the natural order, containing within our bodies a multitude of other organisms and biological systems but also, and sadly, represent among all creatures the species most destructive of that order currently active on the planet. In that light, John Muir's tough questions continue to ring true:

> Why should man value himself as more than a small part of the one great unit of creation? And what creature of all that the Lord has taken the pains to make is not essential to the completeness of that unit—the cosmos? . . . From the dust of the earth, from the common elementary fund, the Creator has made *Homo sapiens*. From the same material he has made every other

creature, however noxious and insignificant to us. They are earth-born companions and our fellow mortals.[36]

Our place in the wider scheme of things can thus be seen, in the light of faith, as radically paradoxical. On the one hand, we are *fallen* creatures who have yet to recognize in full humility the deep kinship we share with other "earth-born companions" and "fellow mortals." On the other hand, we have been invited through grace to *stand* in dignity and hope before God—indeed with and *in* God. Once again, literary imagination might best enable us to apprehend just how such drastic contraries can coexist. As Shakespeare's Hamlet exclaims, in one famous speech, "What a piece of work is a man, how noble in reason, how infinite in faculties . . . and yet, to me, what is this quintessence of dust?"[37] Compared with God, we are lowly beings, particles of Adamic dust drawn from the earth, but we are at the same time children of God, formed from the luminous residue of stars.

As author Wendell Berry often calls to mind, that which for our time seems most pointedly to express our proper place in the cosmos draws on the familial domesticity of "household" language—an idiom linked to the Greek *oikos*, the root of English words like "ecology" and "economics." We are all members, in other words, of God's one great *household* of Creation on earth. Several of Berry's sabbath poems, for example, celebrate that extended community of being, human and otherwise, to which we belong:

The dark around us, come,
Let us meet here together,
Members one of another,

---

36. John Muir, "Cedar Keys," in *John Muir: Nature Writings*, 826.
37. William Shakespeare, *Hamlet*, ed. Barbara B. Mowat and Paul Werstine (New York: Simon & Schuster Paperbacks, 2012), 101–103.

Here in our holy room.
. . . .
Light, leaf, foot, hand, and wing,
Such order as we know;
One household, high and low,
And all the earth shall sing.[38]

There is longstanding precedent for this vision of our membership in God's grand, multi-species household of Creation. I am reminded of how Anne Bradstreet, a seventeenth-century Puritan New Englander who was the first American colonist to publish a volume of poetry, also produced North America's first sustained nature poem in a versified meditation titled "Contemplations." That work, inspired by a solitary autumnal walk through wilderness terrain near the poet's home in North Andover, Massachusetts, expresses Bradstreet's wonder and gratitude to inhabit an ecosystem shared with all manner of beings spawned by her "great Creator." And when Bradstreet, in a prose account of her life story addressed to her children, confesses to having suffered doubts about God's existence and "the verity of the scriptures," she finds her faith confirmed, above all, by her perceptions of the natural world—"by the wondrous workes that I see, the vast frame of the Heaven and the Earth, the order of all things, night and day, Summer and Winter, Spring and Autumne, the dayly providing for this great household upon the Earth."[39]

As an immigrant to the New World who became the mother of eight surviving children, and who as chief matron was charged with "daily providing" for the domestic functions of a leading New England family, Anne Bradstreet would have

---

38. From an untitled poem cited in Wendell Berry's *A Timbered Choir: The Sabbath Poems 1979–1997* (Washington, DC: Counterpoint, 1998), 52.

39. Anne Bradstreet, from "Contemplations" and "Autobiographical Passages" addressed to "My Dear Children," both cited in *Poems of Anne Bradstreet*, ed. Robert Hutchinson (New York: Dover, 1969), 79–87 and 182.

readily understood what it meant to contribute toward the flourishing of life in a "great household." It seems we are called to fulfill a comparable charge today—but as householders of an earthly dwelling place more expansive and much more vulnerable than anything Bradstreet could have known.

Questions for Discussion and Reflection

1. What glimpses of godliness in creation might you have experienced in the course of your own life?

2. Why do you think the familiar "intelligent design" profile of a Creator-God is either appropriate and useful or seriously misleading? How does this imaging relate to scientific theories of an organic and evolutionary creation?

3. In the light of your reading, what relevance do you imagine Christianity's doctrine of the Holy Trinity bears for the practice of earthcare? How might we affirm that relevance while at the same time advancing interfaith dialogue with those representing other faith traditions or none?

4. How exactly could the human faculty of imagination be seen as relevant to the cause of faith-inspired earthcare? How might our own gifts of imagination and creativity best be enlarged?

5. To see ourselves as human "members" of the larger "household" of God's creation is just one among several metaphors used to describe our proper place in the cosmos. What do you take to be the assets and possible limitations of this metaphor?

CHAPTER 2

"The Time Is Out of Joint"

Green Perspectives on Sin, Salvation, and the Fall

> Fall in love with the wild world, and you are taking the first step toward saving it.
>
> —*Margaret Renkl, journalist*

The Burden of Sin and Fallenness

Who can reasonably doubt that the human animal is, morally speaking and perhaps otherwise, a deeply flawed creature? How exactly might we best interpret this lapse from the best version of what any of us can imagine ourselves to be? Should we interpret it, in the light of some Church traditions, as the product of Original Sin? What does it mean to speak of humankind's "fall" from grace and virtue, figuratively expressed through the Bible's Genesis story of a primal act of disobedience in Eden? And just how, and to what extent, might we understand the rest of earthly creation to have "fallen" likewise, mainly as a consequence of human sin?

These are all weighty questions. Yet I think it fair to conclude at the outset, particularly in today's world following the unspeakable horrors of the last century's Holocaust years, that the reality of humankind's capacity for evil is self-evident, even apart from Christian or other religious teaching. As that deep-diving fiction writer Herman Melville once observed, despite his own religious skepticism, the "great power of blackness" discernible in fellow author Nathaniel Hawthorne "derives its force from its appeals to that Calvinist sense of Innate Depravity and Original Sin from whose visitations, in some shape or other, no deeply thinking mind is always and wholly free."[1] The Psalmist, too, insists that no human can claim to be totally righteous—"no, not one," for "they have all fallen away" (Psalm 53:3). St. Paul points out that any honest process of self-examination will confirm this truth, since he knows from personal experience that at times he "can will what is right, but I cannot do it. For I do not do the good I want, but the evil I do not want is what I do" (Romans 7:18–19).

This truth of human sinfulness is scarcely the whole truth, however, from the standpoint either of biblical teaching or of Church tradition. Other theologies have, for example, challenged St. Augustine's conception of Original Sin as a defect directly inherited from our forebears. Reformation-era claims about the total infirmity of the human will must be questioned as well. The Lord's magnificent creation is, after all, according to the Genesis accounts, declared from the first to be indeed "very good," with human beings having been brought to being expressly "in our image, according to our likeness" (Genesis 1:26, 31). We might as well think of the primal world as defined by an original gift, blessing, or peace, as much as by original sin.[2] We are bound to conceive of the fall as a condition—something we endure, together with the whole creaturely order of things—rather than as an event, or a particular human deed. Yet the larger, paradoxical truth of the matter calls us to blend

1. Herman Melville, "Hawthorne and His Mosses," in *Moby-Dick*, ed., Hershel Parker, 3rd Norton Critical Edition (New York: W. W. Norton and Company, 2018), 549.

2. One such reflection in this vein is that offered in *Original Peace: Restoring God's Creation* (New York: Paulist Press, 1997) by David Burrell and Elena Malits.

cognizance of our nature as fallen creatures, all of us impaired by sin and omnivorous egotism, with the assurance that we are also blessedly made in God's image and redeemed in Christ.

If the familiar image of humanity's "fall" seems overly catastrophic and pessimistic, one can also to view it as emblematic of a failing ripe for divine compassion. Thus Julian of Norwich, in her mystical re-envisioning of humanity's fall from grace, writes of how the Adamic figure of a beloved servant runs so hastily though blindly "to do his Lord's will" that he ends up falling into a ditch. Lying there alone and sorely wounded, he cannot even "turn his face to look upon his loving Lord, which was to him full near,—in Whom is full comfort." This servant's "loving Lord" nonetheless "full tenderly beholdeth him," initiating a larger process of transformation whereby "his falling and his woe, that he hath taken thereby, shall be turned into high and overpassing worship and endless bliss."[3] For Julian he becomes, in fact, a figure not only of primal humanity but also of Christ, the Second Adam, who in his incarnation "falls" into the womb of the Second Eve.

St. Paul affirms that the rest of creation, too, shares in humanity's wounded condition. Suffering from a certain incompleteness and "bondage to decay," the whole creation in his view has been "groaning in labor pains," waiting "in eager longing" for a redemptive liberation to be achieved in and through Christ (Romans 8:19, 21–22). But how precisely we are to envision this fallenness, together with the future redemption of creation, is hard to say. We realize, of course, that nonhumans suffer many ills—including, on the part of sensate animals, physical pain, loss, dislocation, terror, and death—even if they are not capable of sin. We may even dream of a transformed biological order, free of predatory violence, that resembles the "Peaceable Kingdom" prophesied by Isaiah (11:1–9) and often

---

3. Julian of Norwich, *Revelations of Divine Love*, adapted by Dom Roger Hudleston, OSB (Mineola, New York: Dover, 2006), 91–93.

represented pictorially by Quaker artist Edward Hicks. A long-ing for New Creation seems, at any rate, both pervasive in the Hebrew Bible and inherent to our psychic being.

What God's redemption of all creation might ultimately look like, however, I cannot pretend to know. I know only that a Christian faith worthy of the name must presume that God somehow wills to bring to fulfillment not human beings alone but everything God had ever created, sustained, and esteemed as "very good." Moreover, it has become all too apparent in our own day how the entire community of life on earth has suffered dis-ease through the greed and blindness of fallen humanity.

Yet another complicating dimension of fallenness to be taken into account is its paradoxically integral, rather than wholly con-trary, relation to redemption. For Christians, the doctrine of a "for-tunate fall," or *felix culpa*, derives from recognition that through the saving deed of Christ's death and resurrection, humankind's moral standing or nature has been marvelously elevated to a level even above that presumed in its pre-lapsarian state.

This paradoxical principle draws us, too, into the larger life mystery of symbiotic processes reflected in nature. We know that through the course of earth's biological cycles death, dissolution, and decay are perforce conjoined with the emer-gence of new life. Poet Walt Whitman helps us remember that dead bodies make for "good manure."[4] Vedic teaching likewise recognizes the interwoven texture of death and life, affirming that Lord Brahma, as creator, must be seen as more nearly partner than adversary of Lord Shiva, the destroyer. And in American English, that season of the year "when comes that other fall we name the fall"[5] reminds us how autumnal lapses toward death and decay open the way to spring's regeneration.

---

4. Whitman, "Song of Myself," in *Leaves of Grass*, ed. Jerome Loving (New York and Oxford: Oxford University Press, 1998), 77.

5. Robert Frost, "The Ovenbird," in *Robert Frost's Poems*, ed. Louis Unter-meyer (New York: Washington Square Press, 1971), 196.

One of Wendell Berry's early sabbath poems, apparently situated in a forest setting and informed by St. Paul's reflections in Romans 8, beautifully captures this cyclic dynamic rooted in a transformational falling and rising:

What stood will stand, though all be fallen,
The good return that time has stolen.
Though creatures groan in misery,
Their flesh prefigures liberty
To end travail and bring to birth
Their new perfection in new earth. . . .
What stood, whole in every piecemeal
Thing that stood, will stand though all
Fall—field and woods and all in them
Rejoin the primal Sabbath's hymn.[6]

A Green Parable of Fallenness and Redemption

One literary text that enables us to probe more deeply the eco-theological implications of human sinfulness and fallenness is Samuel Taylor Coleridge's *The Rime of the Ancient Mariner* (1798). This classic narrative poem of English Romanticism underscores the role of a faith-inspired imagination in shaping our ecological vision. It also looks forward, beyond its own era, toward that which we might call today a "green ethic," expressing our deep kinship with wild animals and the larger community of God's Creation. To address these points, it helps first to recollect some key events in this ballad tale of sin and repentance, as narrated by Coleridge's well-aged Mariner and centered on the experience the Mariner is at pains to share with others.

---

6. Wendell Berry, *A Timbered Choir: The Sabbath Poems 1979–1997*, 13.

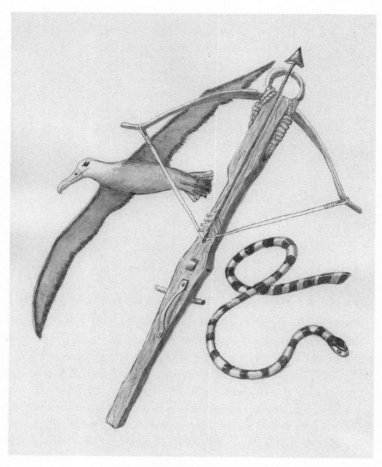

At length did cross an Albatross,
Thorough the fog it came.
As if it had been a Christian soul,
We hailed it in God's name.

I watched the water snakes. . . .
A spring of love gushed from my heart
And I blessed them unaware.
S. T. Coleridge

The Mariner rudely interrupts a wedding guest, on his way toward a joyous bridal feast, to tell his own story of a perilously long journey he had once undertaken at sea. At first the wedding guest tries to repel the intrusion. But he becomes so mesmerized by the bearing of this peculiar stranger, who "holds him with his glittering eye," that he finds he "cannot choose but hear" such a story, apparently meant expressly for him. Doubtless Coleridge hoped his readers, too, might fall under the spell of this tale, with its "strange power of speech"[7] to inspire a vision of how humans and wild creatures inhabit a common earthly *oikos*—in other words, a shared ecological household created by God.

The cross-hemispheric voyage described by the Mariner covers thousands of miles, taking the crew of this sailing vessel through extreme realms of ice, storm, and tropical calm. In the course of these peregrinations, a great white albatross is sighted in the midst of snow-fog and threatening ice floes toward the South Pole. At first the bird is greeted with joy and hospitality. Who, in such a place, could fail to be cheered by the appearance of this large white bird flying on majestic wings? So, "as if it had been a Christian soul," recalls the Mariner, "We hailed it in God's name."[8] As it follows the ship, crew members are pleased to feed it and to call it a good omen. So far, so good.

But then all hell breaks loose. The Mariner confesses to having been abruptly moved to shoot and kill the albatross with his crossbow. He offers no explanation or excuse for doing so. Neither does Coleridge. It seems to have been a thoughtless, impulsive deed, offering no benefit toward

---

7. Samuel Taylor Coleridge, *The Rime of the Ancient Mariner*, in *The Complete Poetical Works of Samuel Taylor Coleridge*, ed. Ernest Hartley Coleridge (Oxford: Oxford University Press, 1966), 187, 208. Subsequent page references to the book, indicated parenthetically in my text, are to this edition.

8. Ibid., 189.

human sustenance or survival. At this point in the story the wedding guest is so struck by what he sees in the Mariner's countenance, even before hearing of the killing itself, that he exclaims with horror, "'God save thee, ancient Mariner! / From the fiends, that plague thee thus!—/ Why look'st thou so?'"[9]

Perhaps we are to suppose that the Mariner killed impulsively out of sheer boredom, by whim or for idle sport— without supposing, anyway, that lordly humans ever needed to justify their slaughter of wild animals. In that light the deed seems emblematic indeed of the "fall," since it seems less an act of deliberate malice than a thoughtless lapse from charity and sound judgment. If to sin is to estrange oneself from God, from one's animate "neighbor" of many possible stripes, and from one's own better self, Coleridge illustrates quite graphically how the Mariner's deed is paradigmatic in this sense as well.

For whatever its motive, the killing is shown here to be not only unwise and irrational but downright sinful in the eyes of God. Sinful enough to provoke dire and even deathly consequences for the Mariner and his crewmates—effects dramatized, symbolically and pictorially, on a supernatural scale. So sinful as to require the Mariner to undergo a long, painful process of personal conviction, purgation, repentance, and confession to a holy hermit before moving toward redemption. Among the trials he must endure is the indignity of having his shipmates hang the dead albatross around his neck. Coleridge portrays these crew members as sharing some of his guilt, to the extent that they had at times actually welcomed what they saw as favorable results of his deed. But the Mariner is, of course, chiefly to blame.

Read in literal terms, without accounting for the figurative hyperbole of Coleridge's storytelling, what the Mariner suffers

---

9. Ibid., 189.

for his crime must strike us as excessive. Does he really, for this one offense, deserve to have each of his crewmates curse him with an eye before expiring? To be haunted by hostile spirits, and by the sight of a skeleton ship, bearing a spectral woman who confronts him with the further nightmare of a Life-in-Death? To be parched with unslaked thirst, unable to sleep, unable to pray? To face "slimy things" that "did crawl with legs / Upon the slimy sea"? And, worst of all, to endure these penalties while fixed in solitary confinement?

> Alone, alone, all alone,
> Alone on a wide wide sea!
> And never a saint took pity on
> My soul in agony.[10]

Yet the principle underscored by his punishment is that wantonly destroying any one of God's sentient creatures violates, in effect, the broader community of all Creation. Coleridge wants us to see the isolation and other dire results of the Mariner's deed as self-determined rather than as simply imposed on him from without. The Mariner's response to the albatross reflects not only disdain for animal life in general but a denial of his own shared involvement in the integrity of creation.

As such, the deed amounts to a failure of imagination—a key term for Coleridge, and one he would in a later remark attach expressly to *Rime*, conceived to be a work of "pure imagination."[11] For as considered in the last chapter, Coleridge understood the Imagination to be that vital, integrative faculty of mind by which we envision the wholeness of reality, the often unseen web of connections that unifies otherwise

---

10. Ibid., 191, 196.
11. From Coleridge's *Table Talk* (1830), cited in *English Romantic Writers*, ed. David Perkins (New York: Harcourt Brace and World, 1967), 404–405.

disparate elements of God's creation. Imagination might thus be seen as an inherently ecological cast of mind, an inward counterpart of the web of life.

The Mariner's dearth of imagination has moral consequences because it leaves him unable and unwilling to perceive any creaturely kinship between himself and the animal he thoughtlessly destroys. He is blind to all he shares spiritually and even biologically (including, as we recognize today, a common preponderance of DNA coding) with the albatross.

But just as the Mariner's sin against an animal, the "neighbor" so mysteriously brought near to him,[12] plunges him into radical isolation and miserable exposure to death-in-life, so also the grace of his return to fellowship with the community of Creation is marked by yet another encounter with animals. In this case, significantly, with animals that most humans find repellent:

Beyond the shadow of the ship,
I watched the water-snakes
They moved in tracks of shining white,
And when they reared, the elfish light
Fell off in hoary flakes.

Within the shadow of the ship
I watched their rich attire:
Blue, glossy green, and velvet black,
They coiled and swam; and every track
Was a flash of golden fire.

---

12. We are of course, enjoined, in the decalogue and in Jesus's second Great Commandment, to love our neighbor as ourselves. And in our own day we are primed to recognize how the resonating question of "Who is my neighbor?" from St. Luke's Parable of the Good Samaritan applies to nonhumans as well. Much of Wendell Berry's prose and poetic writing addresses this theme, thus affirming "neighborliness," as an essential gospel virtue.

O happy living things! No tongue
Their beauty might declare:
A spring of love gushed from my heart,
And I blessed them unaware:
Sure my kind saint took pity on me,
And I blessed them unaware.[13]

It is partly owing to the phosphorescent glow surrounding
these sea snakes by night, we might suppose, that the Mariner now finds them beautiful and is moved to bless them. But
his affirmation of their beauty, stirred by a reawakening under
grace of his imagination, has still more to do with realizing
that they are indeed creatures of God—as is the albatross,
together with beings like himself. The sore afflictions he has
lately endured also seem to have bridled his egotism, opening
him up to receive the grace now offered. Coleridge's prose
annotation at this point of the poem establishes that "By
the light of Moon he beholdeth God's creatures of the great
calm." He sees these creatures now, both literally and figuratively, in a new light. Toward the poem's close, Coleridge
encapsulates the gospel truths about *reverence for life*[14]—to
invoke Albert Schweitzer's telling phrase—with which his
restored Mariner will henceforth evangelize others. These
familiar lines, taken to heart, may become searching rather
than sentimental for us if situated within the narrative events
that precede them:

He prayeth best, who loveth best
All things both great and small;
For the dear God who loveth us,
He made and loveth all.[15]

---

13. Coleridge, *The Rime of the Ancient Mariner*, 198.
14. Albert Schweitzer, *Out of My Life and Thought: An Autobiography*, trans.
C. T. Campion (New York: Holt, 1949), 158–159.
15. Coleridge, *The Rime of the Ancient Mariner*, 209.

The Mariner, despite his original assault on life and fall from grace, thus undergoes a genuine conversion of heart. We might take the particular, twofold character of that conversion as emblematic for our time. It involves not only an awakening to gospel faith but also a commitment to ecological preservation. In the same vein, Pope Francis, in his *Laudato Si'* encyclical, writes of how today's "ecological crisis is also a summons to profound interior conversion." Echoing Pope John Paul II, he underscores the imperative for Christians across the globe who are indifferent or hostile to earthcare to experience an "'ecological conversion,' whereby the effects of their encounter with Jesus Christ become evident in their relationship with the world around them."[16]

The Meaning and Reach of Salvation

The ecological parable represented in *The Rime of the Ancient Mariner* dramatizes the character and destructive consequences of human sin but also offers toward its close a glimpse of salvation: the Mariner's restored personal communion with "all things."[17]

But what, in essence, does "salvation" really mean in the full light of gospel teachings applicable both to us and to "all things"?

I think it worth emphasizing, first, that its meaning should not be shriveled to the narrowly personal, rather self-serving sense it holds in the popular mind. Salvation cannot refer simply to a hope that I will "go to heaven" after I die—ideally, to enjoy there the company of a few other souls I hold dear. It cannot be all about *me*, slight creature that I am. It has to mean more than claiming I've had a "born again" experience.

---

16. Pope John Paul, *Laudato Si': On Care for Our Common Home* (Huntington, Indiana: Our Sunday Visitor, 2015) 8, 141.

17. Coleridge, *Rime of the Ancient Mariner*, 209.

Neither can being "saved" mean that I have arrived at a once-and-forever assurance of having been set apart from others for divine favor. Thus Coleridge's Mariner, even after he is restored to communion with the wider world, must continue to work out his salvation. He remains restless, occasionally anxious, compelled to keep sharing his story because "till my ghastly tale is told" each time "this heart within me burns."[18] No one and no thing can yet be counted as altogether "saved" within the temporal flux of mortal existence.

In biblical settings, the hope of salvation enfolds a richer span of connotations than popular beliefs allow—some pertaining to the individual but others to collective bodies such as the Hebrew people, or to creatures and creation as a whole. Yes, biblical salvation has much to do with our deep desire for liberation from personal sin and death but reaches even beyond that. Salvation aligns with individual pleas for rescue from calamity, with a people's deliverance from slavery or pestilence or other woes, and with humanity's perpetual longing for health and wholeness. It answers likewise to something of an abused land's yearning for restoration and renewal, to creation's own drive toward a *telos* of completion within a Lord who becomes "all in all."

For scriptural authors, especially in the Hebrew Bible, "salvation" (Greek *soteria*; commonly in Hebrew *yeshua*, identified for Christians with the very name of Jesus) often pertains to someone's desperate hope for bodily or other forms of rescue from impending danger. The Psalter abounds in such pleas for deliverance. The Psalmist begs to be rescued from mortal enemies, from illness and plagues, the assaults of wild beasts, famine, flood waters, other natural disasters, and the frightening abyss of death itself. One such invocation has its speaker imploring God to "Save me from all my pursuers,

---

18. Ibid., 208.

and deliver me, or like a lion they will tear me apart; they will drag me away, with no one to rescue me." (7:1–2). Others call for release from bondage to sin, fear, personal failure, spiritual desolation, or despair. The speaker in Psalm 40:2 renders thanks to One who "drew me up from the desolate pit, out of the miry bog, and set my feet upon a rock, making my steps secure." In Psalm 69 the supplicant implores God to "Save me" because "the waters have come up to my neck. I sink in deep mire . . . I am weary with my crying; my throat is parched: and my eyes grow dim with waiting for my God" (69:1–3).

Biblical salvation can likewise be sought, or shown to have been granted already, in collective terms, as characteristically applied in Hebraic thought to the whole people of Israel. The praise sung in Psalm 68, for example, draws on collective rather than individual pronouns:

> Blessed be the Lord,
>> who daily bears us up;
>> God is our salvation.
> Our God is a God of salvation,
>> and to God, the Lord, belongs
>> escape from death. (Ps. 68:19–20)

In fact, the chief manifestation of collective salvation in the Hebrew Bible seems to be the Exodus-event. This defining experience saw an entire people liberated from slavery in Egypt, then to become a nation in a new land. The Lord, speaking to Moses from the burning bush, tells him he has heard the cry of the Hebrews and "has come down to deliver them from the Egyptians, and to bring them up out of that land to a good and broad land, a land flowing with milk and honey" (Exodus 3:7–8). And following the miraculous deliverance by which "the Lord saved Israel that day from the Egyptians," Moses and the Israelites sing together of how "The

Lord is my strength and my might, and he has become my salvation" (Exodus 15:1–2).

A green, largely secular offshoot of these biblical motifs of salvation as rescue can be seen in the form of activist mottos adopted by proponents of the modern environmental movement, especially in the wake of the first Earth Day in 1970, as human-sponsored assaults on biodiversity became more apparent. One such slogan, current through the later twentieth century, set forth an imperative to "save the whales." A signature species for activists, as Gary Snyder's iconic tribute to "Mother Earth: Her Whales" helped to solidify, sperm whales especially were highlighted as vulnerable for extinction because of continued hunting on the part of several nations. John Muir had earlier contributed to this style of promotional evangelism. Even the titles of some Muir essays reflect the impulse, including his call to "Save the Redwoods," and his campaign on behalf of "God's First Temples: How Shall we Preserve our Forests?" Drives to "save" this or that locale or species persist in our own day, of course, with verbal tags of each campaign appearing widely in the form of bumper stickers, bookmarks, placards, posters, dormitory notices, or memes.

While many of these campaigns are targeted toward the rescue of a particular place or life form, broader pleas to "Save the Planet" have also become familiar lately as awareness mounts of a global climate emergency. "Save the Planet" is not a motto to be taken literally, however, in the sense of a rescue from oblivion. The earth is not slated for complete destruction any time soon—not, in fact, until sometime before the dying sun starts to become a red giant, around five billion years from now and eventually swallows up the planet along with much of the rest of our solar system.

Yet current threats to sustaining an appreciable share of the planet's flourishing life forms, together with viable forms of human civilization, remain quite real—and pressing.

Scientists confirm that even if international initiatives to reduce earth's store of greenhouse gasses were shifted promptly into high gear—as seems unlikely, given the current deficit of political will and leadership required—it would already be too late to reverse or even to halt an overall slide toward atmospheric degradation, too late to restore the wondrous old earth of premodern memory. And without some currently unforeseen breakthrough in technology, apparently the best outcome we might expect now is some mitigation of those otherwise grave debasements of planetary life and health that now seem inevitable.

If by "saving" the planet, therefore, we are expecting its restoration to the fullness it once had, that outcome seems hopeless. To preserve as best we can what remains of Earth's beauty, ecological diversity, and capacity for flourishing remains not only possible but also, from the standpoint of biblical values, a moral and spiritual imperative. Yet aside from the Resurrection, even the Bible's many examples of personal and collective rescue fall short of demonstrating that the salvation thus granted is either totally realized or permanent. Following the Babylonian captivity, for example, with the joyous return of Hebrew exiles to Jerusalem, the nation is restored and the Temple rebuilt. But oppressive Roman rule will eventually dash these hopes and destroy the Second Temple. Lazarus, so marvelously raised by Jesus after days in a tomb, still must expect to die eventually.

"Saving the planet," with its undertone of spirituality inherited across centuries of religious culture, is a motto that I believe addresses the health-and-healing aspect of salvation rather than that of wholesale rescue. For beyond all the ills manifested within particular ecosystems or nations, we are confronted today with a specter of climatic dis-ease afflicting the entire globe. That diagnosis is clear, though we remain in search of best prescriptions t[^o] apply toward the patient's health and flourishing.

^ salvation + healing. . .

We find that in the Psalms and elsewhere in the biblical canon, salvation is often associated not only with rescue but also with healing, with the promotion of bodily or psychic health. "O Lord my God," the voice of Psalm 30 exclaims, "I cried to you for help and you have healed me," have even "brought up my soul from Sheol" (30:2–3). In New Testament writings, the many gospel stories focused on Jesus' healing miracles likewise offer signs of salvation in the form of restored health and fullness of life. One such example of an immediately conferred salvation appears in the episode from St. John's gospel where Jesus heals a man born blind. In a revealing gesture of new creation blending mud, saliva, and the divine Word, Jesus enables this man to see as never before. The man then testifies to an inward illumination as well because he knows that Jesus had "opened [his] eyes" to faith (John 9:30).

In other biblical contexts, the reach of salvation extends further still. The speaker in Psalm 36 affirms that as "Your steadfast love, O Lord, extends to the heavens, your faithfulness to the clouds. . . . You save humans and animals alike" (Ps. 36:5–6). And as expounded in Pauline writings and the Book of Revelation, the New Testament's vision of a "new creation" in and through Christ amounts to an all-encompassing salvation. "If anyone is in Christ," writes Paul, "there is a new creation: everything old has passed away; see, everything has become new!" For "in Christ God was reconciling the world to himself" (2 Corinthian 5:17–19).

The "new heaven and a new earth" described in the Book of Revelation likewise urges us to imagine some sort of a radical, cosmic renewal of material reality on the part of One who is "making all things new" (Rev. 21:5). This augury of a universalized salvation is reiterated in chapter 22 of Revelation, where the leaves of the tree of life, situated near God's eternal throne, are said to be "for the healing of the nations."

Among all animal species, human beings exert by far the greatest influence, for good or ill, on discrete land masses, oceans, ecosystems, and in modern times over the planet as an organic whole. Accordingly, initiatives to promote the health of land systems often depend upon processes of moral and spiritual healing taking place within human inhabitants of a given place. Wendell Berry has written, for example, of how he and his wife found this principle playing out in their own lives once they returned to natal ground in Kentucky in 1964 and set out to repair the years of damage and depletion inflicted on the "marginal farm" they purchased there. It was a project of intense, seemingly endless labor but a labor of love. Some of the work, including "the rebuilding of fertility in the depleted hillsides" would take longer than they expected to live. Yet "in doing these things" he discovers that "we have begun a restoration and a healing in ourselves."[19]

That an experience of personal enlightenment or chastening may enable humans to promote, in turn, the health of land and other living creatures is a common feature of the conversion narratives represented in various forms of American environmental literature. Thus John Woolman, a prominent eighteenth-century Quaker known for his antislavery witness, tells in his *Journal* how he had been led toward a conversion of heart after pondering his own mistreatment of animals. He recalls that as a child he had, on a perverse whim, stoned a mother robin and then felt compelled to kill off all her young as well. His conviction of this misdeed so chastened him that "for several hours I could think of little else but the cruelties I had committed and was much troubled." Yet the experience also stirred Woolman to become a better Friend to others,

---

19. Wendell Berry, "The Making of a Marginal Farm," in *Recollected Essays: 1965–1980* (San Francisco: Northpoint Press, 1981), 334.

including animals other than human. "Thus He whose tender deeds are over all His works hath placed a principle in the human mind," he reasons, "which incites to exercise goodness toward every living creature."[20]

Forester Aldo Leopold, in his *Sand County Almanac*, tells a similar tale about how he had once, as an impetuous young man, enjoyed shooting thoughtlessly with friends at a few wolves out West until he found himself gazing at one of those who had been mortally wounded. The sight of "a fierce green fire dying in her eyes" radically transformed his outlook. It led him to appreciate the crucial ecological role of predators and, eventually, toward the deeper wisdom about our relation to nonhuman creation reflected in his celebrated "land ethic."[21]

Several of the Bible's Hebrew prophets remind their hearers that the land's long-term health depends less on the vagaries of "natural" processes than on Israel's willingness to exercise proper husbandry and to hold fast to its covenantal faith, thereby reducing its liability to foreign invasion or exile. Writers throughout the books of Exodus, Leviticus, and Deuteronomy similarly caution that God's gracious gift to Israel of the fertile land they were to inhabit was actually a gift of provisional use, a bounty held in trust under condition of Israel's faithfulness. We, too, would do well to remember that nothing in or on this earth belongs to us absolutely or in perpetuity. The Psalmist affirms that "the earth is the Lord's," together with "all that is in it" (Ps. 24:1), so that we possess nothing from it when we die. Wendell Berry points out, moreover, in his essay titled "The Gift of Good Land," that "the Promised Land is a divine gift to a *fallen* people. For that reason the

---

20. John Woolman, *The Journal and Major Essays of John Woolman*, ed. Philips P. Moulton (New York: Oxford University Press, 1971), 24–25.

21. Aldo Leopold, *A Sand County Almanac and Sketches Here and There* (New York: Oxford University Press, 1987), 130 and 201–210.

giving is more problematical, and the receiving is more conditional and more difficult."[22]

As prophets such as Jeremiah well knew, the consequences of dishonoring this trust could be dire, both for the land and its people. His community's failures had issued in an invasion of Judah from the north, the siege and ruin of Jerusalem, and an ignominious exile in Babylon. Jeremiah is appalled to find the land of promise thus reduced to a scene of desolation:

> My heart is beating wildly; I cannot keep silent; for I hear the sound of the trumpet, the alarm of war. Disaster overtakes disaster, the whole land is laid waste. . . . I looked on the earth, and lo, it was waste and void; and to the heavens, and they had no light. . . . I looked, and lo, there was no one at all, and all the birds of the air had fled. I looked, and lo, the fruitful land was a desert, and all its cities were laid in ruins before the Lord, before his fierce anger. (Jeremiah 4:19–26)

In much the same vein, Joel, prophesying after the exile, would find his community's infidelity responsible for a blighting array of environmental woes spread across the land by a plague of locusts:

> Alas for the day! For the day of the LORD is near, and as destruction from the Almighty it comes. Is not the food cut off before our eyes, joy and gladness from the house of our God? The seed shrivels under the clods, the storehouses are desolate; the granaries are ruined because the grain has failed. How the animals groan! The herds of cattle wander about because there is no pasture for them; even the flocks of sheep are dazed. To you, O Lord, I cry. For fire has devoured

---

22. Wendell Berry, *The Gift of Good Land: Further Essays Cultural and Agricultural* (San Francisco: North Point Press, 1981), 267–275.

the pastures of the wilderness, and flames have burned all the trees of the field. Even the wild animals cry to you because the watercourses are dried up, and fire has devoured the pastures of the wilderness. (Joel 1:15–19)

Torah scribes and other authors of the Hebrew Bible perceived that a land's health, flourishing, and salvation from ruin depended not only on the human community's fidelity to worshipping the one true God but also on its practice of enlightened land management. Such practice included the observance of a land Sabbath, calling for Israel's fields to be left fallow every seventh year. Every fifty years, during a holy Jubilee Year, more radical measures were prescribed to give rest to the soil, to underscore the contingency and transience of all human titles to property, and to ensure a proclamation of "liberty throughout the land to all its inhabitants" (Lev. 25:10). Other stipulations of the Law called for the compassionate treatment of animals. Although these provisions served practical ends, they also underscored the religious principle that the land had a life of its own, worthy of human respect and demanding human restraint, and that all earthly property belonged to God.

According to theologian Ellen Davis, this "abiding interest in land care" on the part of biblical authors reflected an integrative vision of life on earth best described as "agrarianism"—that is, "a way of thinking and ordering life in community that is based on the health of the land and of living creatures." For the writers in question, "the well-being of humans and the enduring fruitfulness of the earth" were conceived to be "inseparable elements of a harmony sometimes imaged as a 'covenant' encompassing all creatures." And while Davis cogently suggests that for these authors "the land's fruitfulness is the 'natural' consequence of covenant faithfulness (*hesed*) enacted on both sides, Israel's and God's," it seems apparent that for them the community's practical observance of Torah stipulations for land management would represent a further dimension of covenant faithfulness. The practical critiques of state-run, commodity-driven forms of agriculture lodged by rurally disposed prophets such as Amos and Hosea offer yet another sign of the culture's "abiding interest in land care."[23]

Neither circumstances leading to the land's degradation and total ruin, nor those contributing to its salvation and restoration, are thus to be attributed solely to some form of supernatural intervention. For the Hebrew prophets, divine agency certainly matters within the ecological ordering of life on earth—but so does human agency, including the practice of responsible land husbandry. That strikes me as an essential principle sustained throughout the Hebrew Bible. It is also a principle applicable to our own circumstance. If abused, the land can often be expected to enact its own penalties to our social well-being, apart from all fancies of divine punishment. Or to cite a relevant chapter title from Rachel Carson's *Silent Spring*, "Nature Fights Back."[24] By the same token, though, biblical authors presume that our species has also been gifted with potential to contribute substantially to the land's health and healing. In that key regard, human beings are indeed called to become co-creators with God toward the flourishing of God's new creation on earth.

---

23. Ellen Davis, *Scripture, Culture, and Agriculture: An Agrarian Reading of the Bible* (Cambridge and New York: Cambridge University Press, 2009), pp. 2, 1, 12, 61, 124–125.

24. Rachel Carson, *Silent Spring* (New York: Fawcett World Library, 1970), 217–231.

A Telling Portrayal of the Land's Salvation

Intended first for children, the *Chronicles of Narnia* by C. S. Lewis is a work of imaginative fiction with much to offer adults as well. It concerns the reach of salvation on several planes, including our perennial yearning to see healing restored to a fallen natural and social order. Within the larger flow of his seven-part series, Lewis suggests how this restoration is linked to the dynamics of the world's genesis as well as to the emergence of God's new creation.

For while Lewis resisted allegorical interpretation of his Narnia tales, Christian imagery and values plainly underlie the series. Rowan Williams confirms that there is indeed "a strong, coherent spiritual and theological vision shaping all the stories."[25] This point is most apparent in the Christological bearing of Lewis's leading character of Aslan, a wild lion who radiates numinous powers of creation, redemption, forgiveness, and love. In several respects, too, Lewis's fantasy country of Narnia is not an altogether alien realm but comes across as a kind of figuratively conceived, parallel universe to our own world. Now and then it is even apt to remind readers of England. Like the world we already know, Narnia had once found its genesis through an outpouring of divine will and love. And like the world we know, its social order is decidedly fallen—tainted by malign spirits, human and otherwise. Its natural order is likewise shown to be disordered, groaning to become more fully what it was meant and longing to be. As portrayed in *The Lion, the Witch and the Wardrobe*, the best-known narrative in the series and the main focus of my remarks, even the seasonal climate rhythms of this land are sadly, to cite Prince Hamlet, "out of joint." Toward the start of *The Lion, the Witch and the Wardrobe*, Narnia suffers under the

---

25. Rowan Williams, *The Lion's World: A Journey into the Heart of Narnia* (New York and Oxford: Oxford University Press, 2012), 4.

malignant and deceitful rule of Jadis, the White Witch who has managed to enlist for her evil designs sundry creatures of this land. As Mr. Tumnus, a Faun, points out, "'Even some of the trees are on her side.'"[26]

At its genesis, however, Narnia had been suffused with the goodness of godly creation. I have noted earlier how, in *The Magician's Nephew* prequel to the Chronicles, Aslan's voice becomes wild but beautiful song, as he joins with other voices to sing the land of Narnia into existence. The larger narrative's re-envisioned genesis story also underscores the goodness of many other nonhuman animals, portrayed as more nearly kindred than as radically unlike the land's human residents. "'Creatures,'" Aslan declares to these other mortals, "I give you yourselves," and "'I give to you forever this land of Narnia.'"[27] Many though not all animals in Lewis's fantasy are shown to be gifted, in fact, with speech and conscious intelligence. And in addition to earth's usual roster of zoological life, Narnia abounds in other species of "wood people" including Dryads, Naiads, Dwarves, Centaurs, and Fauns.

Though Lewis manifested a keen, lifelong appreciation of animal life, he had in earlier years expressed a distinct favor for domestic animals over wilder beasts.[28] But by 1950, when he published the first of his Narnia volumes, he had for

---

26. C. S. Lewis, *The Lion, the Witch and the Wardrobe* (New York: Harper-Collins, 1978), 21. Subsequent page references to the book, usually indicated parenthetically in my text, are to this edition.

27. C. S. Lewis, *The Magician's Nephew* (New York: HarperCollins, 1983), 126, 128.

28. I discuss some of the probable circumstances leading to Lewis's shifting attitudes toward wild animals, including his interaction with the author and spiritual guide Evelyn Underhill, in "'Not a tame lion': Animal Compassion and the Ecotheology of Human Imagination in Four Anglican Thinkers," in *Ecotheology and the Humanities: An Interdisciplinary Approach to Understanding the Divine and Nature*, ed. Melissa J. Brotton (Lanham, Maryland: Rowman & Littlefield, 2016), 197–207.

several reasons been drawn to recognize—as had the Bible's Job-author before him—the distinctive holiness of animal wildness. For as Wendell Berry would later affirm, God "is the wildest being in existence."[29] In Aslan, Lewis had imagined the exceptional case of an animal character who was at once untamed and unfettered, yet decidedly godly. As one Mr. Beaver reminds Earth's four child visitors to Narnia of Aslan's disposition, "'He'll be coming and going. . . . He's wild, you know. Not like a *tame* lion.'" Beaver had likewise responded to child Lucy's earlier query about whether Aslan was "safe" by explaining, "'Course he isn't safe. But he's good. He's the King, I tell you.'" Aslan is in fact "the Lord of the whole wood."[30]

This King's agency, by virtue of a sacrificial death together with Aslan's access to a "deeper magic from before the dawn of time,"[31] will eventually contribute much toward saving Narnia from the false, corrupting sovereignty of Queen Jadis. Yet the obstacles to this restoration are formidable. From the start of *The Lion, the Witch and the Wardrobe*, the land of Narnia had long been fixed in a degraded state of isolation and alienation. Neither do the story's child wayfarers from England at first offer much promise of rescue. In fact, one of these characters, Edmund, only adds to the "fallen" moral atmosphere of Narnia when he betrays his siblings to the cause of Queen Jadis out of selfishness and greed.

In outward, environmental terms, what best epitomizes the lapsed condition of Narnia is its perpetual frigidity. For a hundred years, the White Witch had been able to arrest Narnia's seasonal cycle so that winter never ends—or, as Mr. Tumnus laments, so that it is "always winter and never

---

29. Wendell Berry, "Christianity and the Survival of Creation," in *Sex, Economy, Freedom & Community* (New York: Pantheon, 1993), 101.

30. C. S. Lewis, *The Lion, the Witch and the Wardrobe*, 182, 80, 78.

31. Ibid., 182.

Christmas."[32] Narnia suffers, in other words, from the disruptive and disordering effects of climate change. This is not, of course, change of the "global warming" sort familiar to us today. But it is an unsettling disruption of normalcy nonetheless, one that dramatizes that larger state of ecological, moral, and spiritual crisis in which "the time is out of joint."[33]

What, though, are we to make of the reminder that under the White Witch it is "never Christmas" in Narnia? By some readings, this remark strikes a false note, intruding upon the pagan atmosphere otherwise presumed to define Narnia and undercutting Lewis's disavowal of allegorical intentions. The sentiment seems all the more intrusive once Father Christmas suddenly appears on the scene in Chapter 10 as a figure beaming with jollity, clothed in a "bright red robe," and mounted on a sledge drawn by reindeer with bells. Seen in broader perspective, this interruption with its veiled reference to Christ's nativity is apt. I take it to be dramatizing the ways in which God's new creation often breaks in, struggles to be born, even in a pagan place and time. Grace happens, even in cultural settings beyond all knowledge of Abrahamic faiths. So perhaps it is fitting that along with chosen witnesses in the story, we, too, should be startled, much as Lewis had reportedly been "surprised by joy" in the course of his own life story, by the sight of Father Christmas in a place and time where he doesn't seem to belong. His appearance signals, in turn, the advent of another's saving presence, whose return will restore ecological integrity to the land in accord with the "old rhyme" that Beaver intones:

> Wrong will be right, when Aslan comes in sight,
> At the sound of his roar, sorrows will be no more,
> When he bares his teeth, winter meets its death,

32. Ibid., 19.
33. William Shakespeare, *Hamlet*, 69.

And when he shakes his mane, we shall have spring again.[34]

Aslan does not, however, bring about this restoration single-handedly. Lewis's tale illustrates in several ways the need for humanity to exercise a co-creative agency with God. In the great battle with evil, for instance, which is physically enacted in the penultimate chapter of *The Lion, the Witch and the Wardrobe*, "Peter and Edmund and all the rest of Aslan's army"[35] are expected to join forces with Aslan to quell the White Witch's army.

Lewis affirms most clearly humankind's cooperative role in ushering in the new order through the rituals of royal investment enacted in his closing chapter for those four children from earth. True, the regal imagery surrounding this dénouement may not strike us as compelling or relevant in our present-day setting, especially for Americans, who dispensed with monarchy centuries ago. Why, after all, should we care to learn about the elevation to title rule in Narnia of King Peter the Magnificent, Queen Susan the Gentle, King Edmund the Just, and Queen Lucy the Valiant?

For Lewis, however, monarchy became a defining principle of *The Lion, the Witch and the Wardrobe*, linked to his faith in a divine sovereignty that assured universal meaning and goodness while defying chaos and evil. Michael Ward offers strong evidence that Lewis, in composing his seven Narnia volumes, was inspired in each case by a particular planet—and its mythological aura—from the medieval system. Ward points out that Jupiter, with his reputation for kingship and jollity, was the planetary figure most determinative in Lewis's shaping of *The Lion, the Witch and the Wardrobe*. This identification is relevant because "Jupiter was the king of the planets,

---

34. C. S. Lewis, *The Lion, the Witch and the Wardrobe*, 106, 79.
35. Ibid., 175.

the sovereign of the seven heavens." Alan Jacobs even contends that what the entire Narnia series is all about, perhaps more than anything else, is the issue of "disputed sovereignty."[36]

The crowning of Peter, Susan, Edmund, and Lucy demonstrates how Aslan, as supreme King, despite holding jurisdiction over the entire land of Faery, has nonetheless granted humans substantial though contingent sovereignty for all decisions bearing on leadership and management within the land of Narnia. It is they, and by extension our own species within the contemporary world, who must be held most directly responsible for tending the earth they inhabit. It is they who must hope to be remembered ages hence for having "made good laws and kept the peace and saved good trees from being unnecessarily cut down."[37]

Questions for Discussion and Reflection

1. What do you take to be the true origin and character of sin? Of evil?

2. In what sense might you understand human sin to have infected the rest of creation on earth? How might you reconcile the idea of a "fallen" creation with the affirmations in Genesis of its inherent goodness?

3. There is some question about how the "ecological conversion" that Pope Francis advocates, and that Coleridge's tale of the Ancient Mariner illustrates, can be brought about. What do the you think might be said or done to help stir

---

36. Michael Ward, *The Narnia Code: C. S. Lewis and the Secret of the Seven Heavens* (Carol Stream, Illinois: Tyndale House, 2010), 4; Alan Jacobs, "The Chronicles of Narnia," in *The Cambridge Companion to C. S. Lewis*, ed. Robert MacSwain and Michael Ward (Cambridge and New York: Cambridge University Press, 2010), 274.

37. Lewis, *The Lion, the Witch and the Wardrobe*, 183.

such a conversion, particularly with parties betraying a suspicion of "environmentalism"?

4. Can you think of any present-day cases where a "salvation" of land or nonhuman creatures has occurred, both practically speaking and with spiritual resonance?

CHAPTER 3

Green Perspectives on Christ's Incarnation, the Cross, and Resurrection

> Gather up the fragments left over, so that nothing may be lost.
>
> *(John 6:12)*

The Earthiness of the Incarnation

Even within the largely secular world of present-day Euro-American culture, the festival of Christ's Nativity continues to be celebrated each year in one form or another. For many, it remains at least a joyous occasion of hope, fellowship, good will, and good cheer. And for an appreciable remnant of us, it dramatizes a key tenet of our faith: the Incarnation of God-in-Christ. Despite the sentimentality attached to many of our Christmas customs, this holiday testifies to a great and saving mystery: that the Creator of all things should have consented to take flesh as a Creature, to dwell for a time among us humans on earth, and to alter definitively the relation between Creator and Creation. The sacred mystery we name by that formal-sounding, Latinate idiom of "incarnation," is thus the earthiest of Christian doctrines. It brings infinitude squarely into the homely, mortal and bodily, gutsy stuff of ordinary life.

Other faith traditions, too, suppose that some materialization of godly presence, in the form of avatars, might be discernible here on earth. In Hindu teaching, deities such as Vishnu or Krishna are thought to descend to our temporal world as avatars. Jewish tradition places no credence in avatars but is inclined to recognize a palpable, indwelling presence, or *Shekinah*, suffused throughout God's Creation.

Christian belief in Christ's Incarnation rests squarely, however, on the claim of an actual birth-and-life event situated in historical time. In that crucial respect it differs from the more mythically construed concept of an avatar. As former atheist C. S. Lewis came to believe, especially as influenced by his interchange with J. R. R. Tolkien, the Christ story fulfilled all that had intrigued him about the old pagan myths about dying and reviving gods *but* with the added feature that it was a singularly "true myth," a story rooted in historical fact.[1] It's indeed an extraordinary thought, though admittedly a challenge to faith, to imagine that God-in-Christ once walked, ate, bled, and sweat like any of us, on this very earth. The earthy particularity of that human life seems critical to establishing how God's timeless Word, through whom "all things came into being," entered time and "became flesh and lived among us" (John 1:3, 14). We cannot presume to explain, however, just why this enfleshment took the form it did in salvation history. Just why the Incarnation settled at this discrete time upon a human being from the

---

1. Humphrey Carpenter, *The Inklings: C. S. Lewis, J. R. R. Tolkien, Charles Williams, and their Friends* (Boston: Houghton Mifflin, 1979), 42–45; Michael Ward, "On Suffering," in *The Cambridge Companion to C. S. Lewis*, ed. Robert MacSwain and Michael Ward (Cambridge and New York: Cambridge University Press, 2010), 204–206.

ethnic complexion, gender, and geographic placement of Jesus of Nazareth remains a mystery beyond our ken.

Deep Incarnation

Yet for all its earthiness and rootedness in time and space, the Incarnation's meaning is broadly encompassing as well. From the standpoint of Christian belief, this doctrine's impact in time extends far beyond the earthly lifespan of Jesus. Far beyond the landscapes of ancient Palestine. In our own day, this expansive and even cosmic dimension of gospel teaching has gained new emphasis, sometimes under the title of "deep incarnation." To speak of deep incarnation is to recognize how the Word's becoming flesh qualifies as a uniquely "cosmic event."[2] It is to perceive the incarnational mystery "within the context of an evolutionary world," and as implicated with the entire natural order.[3]

Yet the underlying principle of deep incarnation is scarcely a modern innovation. It can be found within a number of ancient biblical and credal texts. Consider, for example, the famous Prologue to John's Gospel (1:14): "And the Word became flesh and lived among us, and we have seen his glory, the glory as of a father's only son, full of grace and truth." That God assumed a human nature, in the person of Jesus of Nazareth, is indeed a key affirmation that the Prologue goes on to elaborate. The ancient hymn preserved in Paul's epistle to the Philippians sharpens that meaning of

---

2. David Clough, *On Animals, Volume One of Systematic Theology* (London: T&T Clark, 2012), 86–89. For a useful account of the origins and further implications of "deep ecology" language, especially in relation to Athanasius's theology of the Incarnation, see Denis Edwards, *Partaking of God: Trinity, Evolution, and Ecology* (Collegeville, Minnesota: Liturgical Press, 2014), 54–67; and Elizabeth A. Johnson, *Creation and the Cross: The Mercy of God for a Planet in Peril* (Maryknoll, New York: Orbis, 2018), 158–194.

3. Denis Edwards, *Partaking of God*, 54.

incarnation even further. There we are told how God's self-emptying of divine prerogatives in the person of Christ was so complete as to relegate Christ first to the abased status of a human slave and finally to that of a crucified and lifeless body (Philippians 2: 6–8).

It is also significant—though less apparent—that the core linguistic sense of the Johannine "Word became flesh" is simply that the divine Creator became a Creature. To take "flesh" (Greek *sarx*, Hebrew *basar*) means entering the realm of creaturely *animality* and materiality—but not necessarily or exclusively that of *humanity*. Both the Greek and Hebrew refer not exclusively to the bodily stuff of humans but to that of animals as well.

This point is echoed in the two-stage language of a crucial statement in the Church's Nicene Creed, according to which God the Son "became incarnate from the Virgin Mary, *and* was made man" (emphasis supplied). To become a human being thus follows from a deliberately specified antecedent: What comes first, before the species restriction, is the more generically defined sense of "incarnation," whereby a divine being assumes more broadly the attributes of a fleshly and mortal creature. It follows that one does not have to be human to be incarnate.

Undoubtedly Christ's human nature is significant, as Church tradition has always maintained. To what extent, though, should we also consider how the Creator-God's having become a creature, rather than necessarily and solely a human creature, is a critical dimension of the Incarnation? And how might such musings relate to the common perception, shared by members of many faith traditions or none, that certain signs of divine presence are indeed suffused throughout God's creation—though not in the same way, or with the same consequences as the singular Christ-event? The notion of "deep incarnation," coined by Danish theologian Niels

Gregerson, derives from our current need to address questions like these.

As we have seen in the Narnia fictions, for example, C. S. Lewis engages these questions through the audacious stroke of presenting us figuratively with a Christly presence who takes flesh in a nonhuman animal. Lewis portrays the pivotal figure of Aslan as a wild beast—untamed yet clearly benign. Throughout the larger narrative, Aslan enacts the Christ-event not only through his teaching, healing, sacrificial death, and resurrection but also through his linkage, as the Logos or creative Word of God, with a "deeper magic from before the dawn of time." Breathing life into others, he also displays the creative and regenerative force of divine Spirit.

Other texts, too, can further our appreciation of the material and creaturely consequences of the Word's becoming Flesh, thereby drawing us into the very heart of "deep incarnation." One such text is the Latin liturgical chant, "O Magnum Mysterium." A prayer of uncertain medieval origin found in the Christmas Matins service of Roman liturgy, it has been set to music over the centuries by many choral composers, including Tomás Luis de Victoria and Morten Lauridsen in our own day. Particularly when heightened through these musical settings, "O Magnum Mysterium" can awaken us to the rich interfusion of human, more-than-human, and other-than-human feelings aroused by the incarnation event. With heartfelt simplicity, this poetic text conveys the poignant yet all-embracing relevance of incarnational theology:

> O magnum mysterium, et admirabile sacramentum,
> ut animalia viderent Dominum natum,
> jacentem in praesepio:
> O Beata Virgo, cujus viscera meruerunt
> portare Dominum Christum. Alleluia!

O great mystery and wondrous sacrament,
that animals should see the newborn Lord
lying in a manger!
Blessed is the Virgin whose womb was worthy
to bear the Lord Christ. Alleluia![4]

Although the biblical nativity narratives do not tell us explicitly that nonhuman animals attended the birth of Jesus, the mention in St. Luke's gospel of the newborn child's having been laid outside usual human lodgings in a rustic "manger"—that is, a feeding trough for animals—lends support to this familiar imaging of the scene. Tradition added an ox and donkey, both mentioned in the opening chapter of Isaiah. Aside from the holy family, then, the only presumed witnesses to this extraordinary birth would have been domestic stable animals—*not* the distinguished human figures of exalted rank one might otherwise have supposed.

Our text imagines these humble creatures, likely including sheep and oxen, not only as standing before the event but as actively observing its drama. They apparently *watch* what is happening before them. And as these friendly beasts witness another earthly being's messy but marvelous entry to life, might some even register a recognition of having seen comparable births among their own kind? In any case, the beauty and solidarity expressed through their gazing upon Christ surely contributes here to our larger sense of the "mystery and wondrous sacrament" manifested in this birth. Their gazing underscores the ways in which this birth transcends not only the usual species boundaries, but also the great divide between human and nonhuman beings. For that matter, it bridges the still greater divide between Creator and Creation, between

---

4. Latin text cited from *The Liber Usualis*, ed. Benedictines of Solesmes (Tornai, Belgium: Society of St. John the Evangelist, Desclée & Cie: 1956), 382; with my rendering of the translation.

things ordinary and extraordinary, earth and heaven, time and eternity. The "mystery" perceived in this event is nothing like a puzzle to be solved, therefore, but an endlessly expansive play of meanings radiating from what a fifteenth-century English carol calls the "little space"[5] of Mary's womb that distills the presence of all heaven and earth.

Jesus' birth is likewise, by this account, a wondrous "sacrament." In that regard, it is worth recalling that in Christian tradition a "sacrament" has long been defined as a material, outward sign of an inward and spiritual grace. Performing sacramental baptism, for example, requires the material element of water. The sacrament and sacred mystery of Jesus' birth is likewise interwoven with its gutsy physicality, the blood and mess that clearly define this event as a material, as well as a spiritual, phenomenon.

Mary of Nazareth, too, clearly plays a bodily role to enact the enfleshment of this material sacrament, in an integral yet decidedly visceral and even painful fashion. She whose womb was worthy to bear Christ was also destined to bear all the agonies of childbirth—as well as, in union with her Son, the later anguish of his crucifixion and death. Even Christ's holy birth, like the birth of every other human and nonhuman being, carries within it the eventual sentence of death. So does every star and galaxy that ever was or will be. That, too, is a truth of existence enfolded within the theology of deep incarnation.

Deep incarnation must embody not only the birth and life of Jesus but also, finally, the whole of creaturely existence—including death and the cross. To that last point I want to cite a few remarks from Morten Lauridsen detailing how he shaped his own brilliant choral setting of "O Magnum Mysterium." Lauridsen recalls how the text's reference

---

5. The carol in question is best known as "There Is No Rose of Such Virtue."

to Mary, "Beata Virgo," led him to ponder not only the joy flowing from this incarnational birth (and, we might presume, from what the animals saw of it), but also the profound grief Christ's mother must later feel upon seeing her son brutally killed. Accordingly, where the text sounds the Latin word "Virgo," he inserted into the alto voice line a single but telling note of dissonance, thus casting a "sonic spotlight" on what this Mother of Sorrows must eventually face.[6] That touch of dissonance, tempering the music of a joyous birth, strikes me as an apt reminder that the cross, too, falls well within the scope of deep incarnation.

The Cross

As we all know but often want to forget, the actual wooden cross on which Jesus died was a death-dealing instrument of torture and shame. No mode of capital punishment practiced by ancient Rome was considered more abhorrent than crucifixion. Can it make sense, then, that this brutal form of execution should have become an emblem of faith and Christian identity? Should we really be choosing the cross to adorn our homes, prayer books, houses of worship, and bodies?

To this we can dare to say "Yes!"—but only because the New Testament writers insist that the anointed One of God had, indeed, through his willing acceptance of this death, forever transformed, from within, the cross's meaning. St. Paul, recalling a provision of the Law of Moses that "cursed is everyone who hangs on a tree" (Galatians 3:13), points out that Christ liberated the world "by becoming a curse for us." Instead of denying the inherent shame of dying on a tree, Jesus on the cross embraced and thereby transformed that shame.

---

6. Morten Lauridsen, "It's a Still Life That Runs Deep," *Wall Street Journal*, February 21, 2009, W7.

Henceforth, the cross could properly be viewed as a token of honor, triumph, and redemption.

I believe we have yet to understand just *how* the passion and death of Jesus contributed to this liberation for humanity at large and, arguably, for the entire creaturely world. Throughout Church history various theoretical explanations of this "atonement," as it is commonly called, have been proposed. Some of these theories emphasize the ways in which Jesus's death presumably offered a payback "satisfaction" to God's offended honor, the restoration of a cosmic justice that had been fractured by human sin. Other theories would have us adopt a "penal substitutionary view" of the cross, perhaps including the notion that Jesus' vicarious sacrifice on our behalf fulfilled some sort of legal and punitive transaction. In a more popular vein, Jesus's death has even been construed as a form of propitiation, offered to appease a wrathful God.

It makes sense, however, to contemplate the saving work of atonement not solely with respect to the cross but as defining Christ's work of salvation within the whole unified mystery of his incarnation, life, death, and resurrection. St. Paul, declaring how "in Christ God was reconciling the world to himself" (2 Corinthians 5:19), provides forceful support for this view. We might accordingly suppose, in broad terms, that Christ's self-abnegating gift of himself upon—yet also beyond—the cross was efficacious simply by way of reversing the world's otherwise inevitable course of degradation into sin, death, and adversity. Cynthia Crysdale sums up this larger sense of atonement as involving a "reconciliation of estranged parties" so as to bring "the created order back into relationship with its Creator."[7]

---

7. Cynthia W. Crysdale, *Transformed Lives: Making Sense of Atonement Today* (New York: Seabury Books, 2016), 2. This book also provides an illuminating survey and assessment of competing atonement theories across time. Its treatment of St. Anselm's theories (2, 4, 72–90), often misrepresented, is especially discerning.

But even with limited respect to the cross's role in achieving at-one-ment, those familiar transactional or propitiatory theories invite resistance on at least two major points. One has to do with the nature of God; the other, with the scope of salvation. To begin, there is good reason to question whether a wrathful God who requires sacrificial and violent propitiation is really worthy of our belief. Such a deity cannot readily be identified with Scripture's prevailing image through both testaments of a merciful, loving healer who creates and sustains all things. Scripture, tradition, and reason also give us ample grounds for believing that the God of our faith means to extend the ultimate scope of redemption far beyond the case of our own species.[8] Today, more than ever, it seems clear that our faith in God's becoming "all in all" calls us to envision a salvation immeasurably broader than that applied to humanity. With all-surpassing love, the arms of Jesus on the cross reach out—beyond his own immediate torment, and beyond any bounds we can readily imagine.

We might even find hints of this larger reach in the sacred geometry of the cross pattern itself. The broadly encompassing vertical and horizontal planes of this figure extend toward all four coordinates of three-dimensional space. And of course the two planes also intersect—the "crux" of the matter—at a central point. The pattern of the Celtic cross is especially revealing in this regard. For as spiritual guide Esther de Waal has noted, its image of a circle imposed on intersecting lines is "a powerful one," confirming the "interconnection of redemption and creation, that we cannot have the light without the dark." Particularly as displayed in the imposing form of Ireland's

---

8. Such is the major argument sustained throughout Elizabeth Johnson's illuminating book, *Creation and the Cross.* Johnson makes a compelling case for interpreting "the cross and resurrection of Jesus Christ so as to include the full flourishing of all creation" (xvi).

high crosses, it sets before our gaze "the great O of creation, the circle of the world, and the cross of redemption brought together into one whole."[9]

In its fullest reach, Jesus' instrument of shame thus crosses time and space to touch eternity. To this point we can look to the prophetic pronouncements attributed to Jesus in St. John's Gospel. There Jesus speaks of how when he is "lifted up from the earth," he will "draw all people"—as the passage is usually rendered—"to myself" (John 12:32). The elevation in question apparently presages how Jesus must eventually be raised physically onto his cross for execution, even as it points as well toward his resurrection and glorification. But as I recall New Testament scholar Christopher Bryan pointing out in unpublished remarks of his, another quite valid translation of the passage would have Jesus, once raised on the cross, drawing not simply "all people" but "all things" (Greek *panta*) to himself. The statement would thus underscore, once again, the transhuman and cosmic sweep of redemption. Such a reading strikes me as well worth taking into account.

Tree of the Cross

As indicated in the citation from St. Paul in Galatians, Christians had from the first been inclined to think of Jesus' cross as a kind of tree. That figurative association seems understandable, given of course the wooden body and upright positioning of a crucifixion cross. Eventually, too, for Christians reflecting on biblical imagery from both testaments, it became possible to envision the sacred wood of Jesus' cross in conjunction with other biblical trees besides the cursed tree of Deuteronomy 21:22–23. First among these possibilities, Christians

---

9. Esther de Waal, *Every Earthly Blessing: Rediscovering the Celtic Tradition* (Harrisburg, Pennsylvania: Morehouse, 1999), xiii.

envisioned the tree on which Jesus died to be a paradoxical fulfillment of those portrayals, conspicuous in Genesis as well as in the Book of Revelation, of a Tree of Life. They entertained a vision of how, amid a horrid death, new life of several kinds had sprouted from Calvary's crossbeams.

One popular legend of Jesus' cross supposes it came from a Dogwood tree.

The Latin hymn "Pange Lingua Gloriosi" ("Sing, my tongue") from the sixth century gives memorable expression to these themes in words commonly sung during the Good Friday liturgy, as the cross is borne into the church. The text's presumed author, Venantius Fortunatus, underscores not only the triumph of the cross but also its fecundity and its vital engagement—in beauty as well as agony—with our own material world:

Sing, my tongue, the glorious battle;
of the mighty conflict sing;
tell the triumph of the victim,
to his cross thy tribute bring.
Jesus Christ, the world's Redeemer
from that cross now reigns as King.

. . . .

He endures the nails, the spitting,
vinegar, and spear, and reed;
from that holy body broken
blood and water forth proceed:
earth, and stars, and sky, and ocean,
by that flood from stain are freed.

Faithful cross! above all other,
one and only noble tree!
None in foliage, none in blossom,
none in fruit thy peer may be:
sweetest wood and sweetest iron!
sweetest weight is hung on thee.

Bend thy boughs, O tree of glory!
Thy relaxing sinews bend;
for a while the ancient rigor
that thy birth bestowed, suspend;
and the King of heavenly beauty
gently on thine arms extend.[10]

In this reflection, Christ's "tree of glory" is envisioned as combining, within its organic presence, the life-force that sustains God's ongoing work both of redemption and of cosmic

---

10. Venantius Honorius Fortunatus, trans. after John Mason Neale, cited from *The Hymnal 1982* (New York: Church Hymnal Corporation, 1985), 165–166.

creation. Accordingly, the "sweetest wood" of this tree of life branches out to connect with its figurative counterparts—again in the Book of Genesis, toward Scripture's start, as well as in the Book of Revelation's tree bearing "twelve kinds of fruit" and leaves "for the healing of the nations" toward Scripture's close (Revelation 22:2). The fuller text of Fortunatus, omitted from most hymnal versions sung today, even invokes legends of how a branch from that other tree, the one bearing forbidden fruit in Eden, might have been planted thereafter at Golgotha—thus reinforcing the sense of radical transformation from blighted wood to new creation on the cross. But the most striking feature of this "Pange Lingua," from the standpoint of green faith, must surely be the way it portrays the flow of blood and water from the "holy body broken" to be an outpouring of godly life upon all creaturely existence: earth, stars, sky, and ocean.

This arresting imagery might for us today call to mind how a "tree of life" has also, long after the time of Fortunatus, been adopted by Charles Darwin among others as a biotic figure for the interbranching evolution of varied species and life forms from primordial roots. It might recall, for that matter, Yggdrasil, the sacred tree of cosmic import from Norse mythology. "The affinities of all the beings of the same class have sometimes been represented by a great tree," Darwin observed, whose "green and budding twigs may represent existing species; and those produced during each former year may represent the long succession of extinct species."[11] Doubtless Darwin was not imagining here any identification with the tree on which Jesus died. Yet from the standpoint of faith, we might well perceive a linkage between the vision Fortunatus sets forth and later emblems of an ecological unity-amid-biodiversity. This poetic association may, in addition, deepen

---

11. Charles Darwin, *The Origin of Species: Revised Edition*, introduced and abridged by Philip Appleman (New York: W. W. Norton, 2002), 74.

our awareness of how life and death were forever interfused in the cross of Christ.

Shadows of the Cross in our World

Christians have always understood Jesus' passion and death to be meaningful far beyond the bounds of its time and place in first-century Palestine. Our tradition affirms that Jesus on the cross not only accepted his own ordeal but also took upon himself all the pain, sin, and grief of our species—and, for that matter, of the entire creaturely world. The cross brings saving health. And yet, together with the One fixed upon it, it never ceases to embody as well the immensity of death, affliction, and loss that pervades Creation. As the Man of Sorrows, Jesus willed to become fully engaged with this grief, particularly by virtue of his ministry on behalf of poor, disdained, and disabled human beings.

So also in our own time, shadows of the cross continue to fall especially on certain dispossessed sectors of humanity. Such persons include those whose lives have been ruined by human-induced famine, or toxic environmental incursions, or who belong to an ever-growing throng of climate refugees with no home on earth. The present-day ecojustice movement aims to highlight and address their plight. We also want to recall that St. Paul, in his day, perceived how the "whole creation" had been "groaning in labor pains," showing eagerness for release "from its bondage to decay" (Romans 8:18–23). And today that groaning of our planet is all the more audible, even desperate.

That we have entered lately a new era of the "Anthropocene," in which humans alone exercise *the* decisive influence over the planet's fate, is scarcely cause for pride in our species, though it does allow us certain opportunities. In any case, the litany of woes afflicting our planet is now too long and frightening to justify complacency. We are already, for example, enveloped in a sixth era of global extinction that has seen an accelerating, massive-scale erasure of plant and animal species. It's not only species that are fast disappearing, though. Over the last half century, earth's total population of wild animals has been reduced by roughly half, with about a third fewer North American birds alive than previously. Throughout much of the world, access to clean air and water has already been severely compromised, as has the organic integrity of many croplands and rainforests. Massive deforestation, dying coral reefs, and the rapid depletion of oceanic life forms have yet to be countered effectively.

A climate change "emergency" has plainly arrived now in force. It has already begun to bring, and threatens to bring still more egregiously, all manner of disasters: rising sea levels, floods, more frequent and severe storms, new wars, wildfires, droughts, famine, displaced populations, and the demise of island nations and whole cities. Miami and many other seaside settlements are already on the way to being swallowed up. Feedback loops can be expected to hasten the process. No wonder we have reason to fear we are approaching and may even have reached—to invoke a telling phrase from environmentalist Bill McKibben—"the end of nature."[12] Not, in this sense, the end of all life on the planet but at least a permanent demise of that fertile, vibrant, splendidly adjusted Earth of the Holocene era that many of us once knew. That is horror enough.

Even for those of us sustained by belief in a loving, potent, and beneficent God, it is hard to retain hope for the future in the face of all this. And it would be a grave, prideful error to suppose that God will somehow rescue the planet from degradation apart from, or even despite, our human responses to the

---

12. Bill McKibben, *The End of Nature* (New York: Anchor-Doubleday, 2nd ed., 1999).

crisis before us. Neither can it help simply to become mired in fear and despair over what is happening. Yet by way of deepening our cognizance of all that the community of life stands to suffer and lose within today's shadow of the cross, I think it's of some use to recollect particular places on earth that have for each of us continued to stir our love for the earth and thirst for meaning.

From my childhood years into adulthood, one locale that has long held such meaning for me lies amid the lake country and forested peaks of New York's Adirondack Mountains. In that same state, home places of fond memory include settings in two of its smaller cities: Schenectady and Ithaca. Another land where I once lived but never left in spirit is the picturesque "quiet corner" of northeastern Connecticut, with its rolling hills and historic village greens. The splendor of Europe's Alpine peaks and flowered meadows is heightened all the more in my mind by recollections of having hiked through much of this terrain with my wife and beloved Swiss friends. Closer to home and the present moment, I am grateful for the chance to absorb each day the graces of living in a rural Tennessee college town surrounded by 12,000 acres of mostly unspoiled woodlands, streams, and fields.

Shadows of the cross fall on these places, too, but not yet so far as to darken all the joy they impart to all of us who hold them in memory.

Resurrection of the Dead

More than all other precepts of orthodox belief, or any of Scripture's moral teachings, the Gospel proclamation of Resurrection—that is, of God's having raised up Jesus from the dead as "first fruits" of a general resurrection—stands as *the* very cornerstone of Christian faith. St. Paul could not have driven home this point more emphatically than he did. "How," he asks, "can some of you say there is no resurrection of the dead?" For "if there is no resurrection of the dead, then Christ has not been raised; and if Christ has not been raised, then our proclamation has been in vain and your faith has been in vain" (1 Cor. 15:2–14).

Despite that much clarity on the matter, together with some fine biblical commentary that can clarify still more, the gospel claims about resurrection remain a tough sell. Believing that Jesus rose from the dead does not come easily for many or most of us today. In fact, it never has. Neither does absorbing inwardly, for ourselves, the overarching meaning of this unfathomable mystery. And what should we make of the way both Paul and evangelistic authors of the resurrection narratives underscore the bodily, materially tactile character of the risen Jesus? Especially as portrayed in John's gospel, the Risen One eats fish and bread, invites the apostle Thomas to touch him, and still bears the wounds of his passion. Yet from the first, diverse theories have been proposed to explain away the materiality of this extraordinary case claimed by biblical witnesses—so as to construe the resurrection appearances, if they happened at all, as some sort of disembodied vision or apparition. Thomas has scarcely been alone in doubting the material veracity of what witnesses reported to him.

True, we are bound to notice the many forms of material renewal, of life's ongoing emergence out of death, that prevail throughout the physical world—including the resurgence of plant life that cheers us each spring after winter's darkness. These familiar patterns of dying, decay, and rising again offer us, though, only a distant analogue of what true resurrection is about. For in this world as we know it, all living organisms look to be invariably and implacably mortal. Or in the penetrating words of Hazel Motes, Flannery O'Connor's offbeat fictional character who preaches on behalf of a "Church

Without Christ," it appears certain that "'What's dead stays that way.'"[13]

"What's dead stays that way" looks indeed to be the universal rule of existence, from the standpoint of all we experience objectively around us. But from the interior perspective of our heart's knowing and longing, we can begin to find those gospel claims about resurrection making much more sense. To my mind, that's because they accord so beautifully with our innate conviction of another, eternal plane of existence, with which our destiny is somehow linked, beyond a waking reality bounded by time and space. Stories of Jesus' post-resurrection appearances, wherein he is barely recognized and passes with a transformed body through locked doors, speak directly to this sense of things. Yet the full meaning of the gospel of resurrection continues to elude us. It defies most rational efforts to understand or to explain in depth. In that regard, we might recall how, in St. Mark's Gospel, the women who first gazed into the empty tomb of Jesus and learned that he "ha[d] been raised" (Mark 16:6) did not for a time feel moved to tell anyone anything about their experience. Amazed and afraid, they found no words for what they had encountered. Under the circumstances, though, their silent astonishment may have been a fitting response. We might still learn something useful from that.

But if the full meaning of resurrection must elude us all in this mortal life, we might at least begin to approach its essence indirectly by pondering some common beliefs that I take to be *not* compatible with gospel teaching.

One such belief would equate resurrection of the dead with achieving, through some marvelous means, the medically inexplicable resuscitation of a corpse. The story of Jesus' raising of his friend Lazarus, as related in St. John's

---

13. Flannery O'Connor, *Wise Blood*, in *Three by Flannery O'Connor* (New York: New American Library, 1962), 60.

Gospel, offers a case in point. Lazarus had already been dead for days when Jesus arrives. Yet Jesus assures Martha that "your brother will rise again"—not only "on the last day" (John 11:23–24), which as one sort of Jew she already believes, but closer to the here and now, consistent with that foretaste of the endtime and sharing in eternal life that theologians often call "realized eschatology." Lazarus, responding to a loud summons from Jesus, walks forth from his grave with face, hand, and feet still bound in burial clothes. For this gospel writer the episode stands as a revealing *sign* of God's power in Christ to raise the dead. But it is not itself an instance of resurrection. So far as we know, Lazarus undergoes no bodily or other great change aside from regaining consciousness. And as we surely *do* know, Lazarus once he returns to life must still expect to die, without further resuscitation, sometime in the future. His time of genuine resurrection has yet to come.

Another familiar but largely unbiblical belief centers on the prospect of personal immortality, generally equivalent in Platonic terms to an immortality of the soul. It is a belief commonly held even within Church circles. Those who entertain this expectation may further suppose that for humans death is more of an illusion, an immediate passage to felicity, than a crisis to be mourned or feared. But Christian and Jewish teaching maintains instead that God alone is inherently immortal—that is, never subject to death. And the New Testament's varied accounts of the Christ-event all concur that Jesus, though subsequently raised from the grave, suffered a genuine and decisive death, which he therefore approached with some apprehension.

Belief in the purely immaterial survival of an immortal soul also runs contrary to biblical assurances, particularly enforced in St. Paul's writings, of the *bodily* resurrection initiated by Christ. Throughout their testimonies, New Testament authors are at pains to contrast the physically palpable yet radically transformed and glorified body of the risen Christ from the disembodied soul of a ghost figure. They consistently attest that the body of Christ resurrected displays at once noteworthy *difference from* and essential *continuity with* that of the Jesus who lived before. In the process, they can heighten our awareness of the inherent linkage between spirituality and materiality, as well as of the transformational dynamic germane to our vision of the material world and the prospect of cosmic redemption.

Transformed Bodies and the Redemption of Materiality

A questionable feature of some forms of resurrection evangelism—and, still more, of assurances about the soul's immortality—lies in their supposition that any faith-bolstered belief in "life after death" has mostly, if not entirely, to do with my own fate as an individual. What will happen to me when I die? Or as cruder formulations of the faith have for centuries led many to wonder: Will I go to heaven, and what must I do or believe while alive to get there?

These may be fair questions, as far as they go. But since they are all about *me*, they are in themselves too narrowly self-interested to engage the much greater question posed to us by scriptural revelation in its entirety: What does it mean for God in Christ to have willed, and to have effected in self-abandoning love, the *redemption of the world* that God created? What does it mean that God so loved the world as to die for it? And the "world," in this familiar usage, should probably signify the collective whole of humanity across time, if not indeed finally the whole of cosmic creation. Not just an individualized "me," in other words. A communal rather than individualistic vision of resurrection is already reflected in the

Hebrew Bible, at least by analogy, in Ezekiel's famous prophecy of God's promise to restore the body of Israel to new life, as imaged through a revivified valley of dry bones. Dante, toward the close of his *Divine Comedy*, likewise envisions God's resurrected faithful ones as mutually conjoined members, appearing as floral petals, within the grand, decidedly organic image of a Celestial Rose.

We can scarcely forget that every body on earth, human and otherwise, returns to dust when it dies. Nothing of its identity or integrity remains visible. So the bodily aspect of gospel claims for resurrection has from the first proved hardest for most people to believe. Yet it also provides the strongest support for our faith that God will ultimately redeem, somehow or other, the whole of creation—including time, space, and the material order itself. For as Christopher Bryan points out, the "transformed physicality of Jesus' resurrection" is at the root of this hope. Neither, he attests, "is that vision of hope limited to humanity." Instead, "it is finally cosmic," so as to confirm that "the entire created order shall have a share in its unfolding salvation."[14]

Theologian N. T. Wright concurs, finding Christ's resurrection to be

> the sign of hope for the future, not only for individuals but for the whole world. As Paul saw so clearly in Romans 8, it was the sign that the whole creation would have its exodus, would shake off its corruption and decay, its enslavement to entropy. The New Testament is full of the promise of a world to come in which death itself will be abolished, in which the living God will wipe away all tears. The personal hope for resurrection is located within the larger hope for

---

14. Christopher Bryan, *The Resurrection of the Messiah* (New York: Oxford University Press, 2011), 184, 186, 189.

the renewal of all creation. Take away the bodily resurrection, however, and what are you left with? The
development of private spirituality, leading to a disembodied life after death: the denial of the goodness
of creation, your own body included.[15]

Granted, it is impossible for us to conceive in graphic
terms just what such an ultimate, radical renewal of creation
might look like in the age to come. We can only guess about
the particulars of this transformation. Thus C. S. Lewis,
when drawn to speculate about the possibility of a renewed
existence for at least certain higher animal species in the
new order, met with ridicule from some. Nothing daunted,
though, he declared mischievously that he was not "greatly
moved by jocular inquiries such as 'Where will you put all the
mosquitoes?'—a question to be answered on its own level by
pointing out that, if the worst came to the worst, a heaven
for mosquitoes and a hell for men could be very conveniently
combined."[16]

As regards the issue of bodily resurrection, St. Paul discerned clearly that the very concept of bodiliness can be
variously construed. Human as well as nonhuman bodies,
though grounded in materiality, also possessed a constitutive
form that transcended static being—that even transcended,
in diverse figurative and practical modes, their own individ-
uality. Paul's musings in 1 Corinthians 12 upon our common
participation in more-than-physical bodies, including the

---

15. N. T. Wright, "The Transforming Reality of the Bodily Resurrection," in
Marcus J. Borg and N. T. Wright, *The Meaning of Jesus: Two Visions* (San Francisco: HarperCollins, 1999), 126.

16. C. S. Lewis, *The Problem of Pain* (New York: Macmillan, 1962), 137. For
further exploration of Lewis's views on animal pain, consciousness, and domestic
vs. wild animality, see John Gatta, "'Not a tame lion': Animal Compassion and
the Ecotheology of Human Imagination in Four Anglican Thinkers," in *Ecotheology in the Humanities*, ed. Melissa J. Brotton (Lanham, Maryland: Lexington,
2016), 197–207.

body of faithful believers and the body of Christ, is a case in point. And when he confronts in this same epistle some vexing questions about resurrection of the dead—that is, "How are the dead raised? With what kind of body do they come?" (1 Cor. 15:35–36)—he insists above all on the essential *transformation* of post-mortem bodies that this mystery entails.

Remarks by theologian Christopher Bryan can help us grasp more viscerally what bodily existence and the transformed physicality of resurrectional bodies might mean in a contemporary light:

> Of one thing we may be sure: it is in this world, this *physical* world, where much evil is done. It is *bodies* that are starved and raped and beaten and tortured and abused. It is in their *bodies* that Jews and others were enslaved and murdered at Auschwitz and Buchenwald and similar places. It is the *bodies* of Japanese men, women, and children that were destroyed at Hiroshima and Nagasaki. It is in their *bodies* that African American men and women were humiliated and abused for generations. It is in their *bodies* that living animals are farmed as if they were products off a conveyor belt. It is in its *body* that the good earth is strip-mined and left in desolation. . . . We honor God when we are true to ourselves, and God made us to be, among other things, physical. Moreover, God has honored that physicality not only by pronouncing it "very good" but also by becoming it.[17]

We may suppose our own age of scientific rationalism to be particularly susceptible to doubts about the plausibility of

---

17. Bryan, *The Resurrection of the Messiah*, 185–186.

bodily resurrection. Yet it seems to me that modern science also allows us to perceive, better than anyone alive in the Greco-Roman era, the radical contingency and permeability of all physical bodies. We now understand our own human body to be teeming with micro-organisms, as a protean and dynamic vortex of mutability in which virtually all its constitutive cells will have been exchanged with other cells prior to our death. The human body, regarded either microbiotically or in connection with the macro-processes of critical organs, is far from a solid and static substance. We now know it to be subject, for good or ill, to newly discovered forms of genetic manipulation, and to the surgical exchange of living body parts from other humans or nonhumans. I know that my body bears countless hydrogen atoms that emerged shortly after the dawn of time, and will cycle perpetually into other settings so long as the cosmos endures.

That is not to say that these developments, or any form of scientific or historical evidence, offers us proof for the resurrection of the dead. But they may at least enable us to sense how Paul's assurance that, especially in definitive bodily terms, "we will all be changed" in an age to come need not be considered quite so preposterous as our first, common sense impressions might dictate.

In still larger terms, the redemption of materiality itself plays into our gospel hopes of resurrection. So does the question of what a "new heaven and earth" might mean, given what we now recognize as the ongoing dissolution of the entire physical universe. We know that in some five billion years, for example, our own sun will begin to run out of hydrogen and die, after first expanding into a red giant and swallowing up the inner planets of the solar system.

We also know that, after billions more years, the accelerated dispersal of galaxies will signal that the universe itself is headed toward total extinction. Endings of this sort must now strike us as inconceivable. Our faith tradition nonetheless testifies that, in some future order beyond time itself, the Creator God can and will bring novel things to birth from what is old.[18] As the Creator is said to declare, in the Book of Revelation (21:5) following from Second Isaiah, "See, I am making all things new."

In trying to imagine what it might feel like to experience such an ultimate, transformative resurrection of materiality under grace, I like calling to mind scenes from a luminous novella by Wendell Berry, titled *Remembering*. Its protagonist, Andy Catlett, is a displaced soul whose alienation and despair drives him away from his farming roots in Kentucky to undertake a long, roundabout and painful journey that brings him eventually back to his home ground. But he returns to his all-too-familiar community of origin and life graced to re-envision it in a visionary light, transfigured. It's as though he has learned to apprehend the world in an expansively divine and eschatological perspective—that is, within the frame of an endtime beyond the constraints of ordinary time. Yet within this perspective the place's distinctive material and vitalizing features are not effaced but renewed and transformed. Andy comes home to discover both the settled and natural faces of his natal community freshly revealed to him

> as he never saw or dreamed them, the signs everywhere upon them of the care of a longer love than any who have lived there have ever imagined. The houses are clean and white, and great trees stand among them and spread over them. . . . Over town and fields the one great song sings . . . and in the

---

18. Such prospects of the distant future are more fully explored in Christian perspective by astrophysicist Arnold Benz in *The Future of the Universe: Chance, Chaos, God?* (New York: Continuum, 2000).

fields and the town, walking, standing, or sitting under the trees, resting and talking together in the peace of a sabbath profound and bright, are people of such beauty that he weeps to see them. . . . He sees that they are the dead, and they are alive. He sees that he lives in eternity as he lives in time, and nothing is lost.[19]

Nothing is lost. Despite the countless deaths and diminishments to which all creaturely life is subject, that may be the best motto available to express the otherwise inexpressible gospel of a resurrectional hope embracing all creation. One version of the gospel story about Jesus' miraculous feeding of a large crowd with barley loaves and two fish likewise concludes with Jesus exhorting his disciples to "Gather up the fragments left over, so that nothing may be lost" (John 6:12).

It would be a mistake, though, to equate resurrection of the dead with a facile "pie in the sky" optimism dismissive of life's tragic realities. In the face of those realities, and mindful of our inevitable deaths, we could do worse, perhaps, than to fall back upon a simple trust and acceptance of divine mercy for all that lies ahead. It is a disposition beautifully portrayed by poet Jane Kenyon, in lines counseling herself and us to "Let Evening Come." Here she envisions how the end of each day unfolds with fading light over the cricket, the fox, commonplace features of the workaday world, and her own person. And foreshadowed, of course, with the end of each day, comes the end of each existence, and the eclipse of most everything we humans can know or imagine. Yet, with a candor most anyone must find disarming, Kenyon voices a calmly assured response to all this:

---

19. Wendell Berry, *Remembering* (Berkeley: Counterpoint, 2008), 102.

Let it come, as it will, and don't
be afraid. God does not leave us
comfortless, so let evening come.[20]

Questions for Discussion and Reflection

1. Why in particular might the Church's traditional affirmation of God's "incarnation" in Christ matter, for us and the rest of the created order? What difference does this teaching make, compared with other views of Jesus as simply a great moral teacher or prophet?

2. Only in recent years, and only in certain quarters, has it become usual to speak of "deep" incarnation. But how might this view of God's involvement with the world be seen as implicit in the original gospel proclamation?

3. Sometimes one hears that preaching a message of "Christ crucified on the cross" conveys all or most of what's essential about the gospel proclamation. What might be said on behalf of this outlook? What though, does it fail to recognize or include?

4. In hymnody and other forms of imaginative reflection found in Christian tradition, the cross upon which Jesus died has commonly been described as a kind of living tree. What significance does this figurative association hold for green faith? And if our own engagement with the cross isn't restricted in time to its first-century use as an instrument of torture and execution, what does it mean to live under the cross in our own day?

5. Why do you suppose Paul and other New Testament authors regard Jesus' resurrection from the dead as the

---

20. Jane Kenyon, "Let Evening Come," in *Otherwise: New & Selected Poems* (Saint Paul, Minnesota: Graywolf Press, 1996), 176.

indispensable cornerstone of Christian belief? What implications beyond our own fate as individuals should we recognize in this teaching?

6. What parallels does "resurrection of the dead" bring to mind for us with patterns of birth, growth, death, dissolution, and rebirth we see played out perpetually within the natural order? What, though, is the crucial difference between the resurrectional proclamation and these biotic cycles of birth, death, and rebirth?

CHAPTER 4

Toward a Gospel of Hope, on and for Earth as It Is in Heaven

Practicing an Earth-Hearted Hope

It is hard to remain hopeful as we witness the dispiriting degradation of the planet now taking place. The health of our common home looks more doubtful today than it has since the dawn of human reckoning. It is hard to hope, too, in the face of humankind's perennial failure to fulfill its own dreams of ensuring a just and flourishing social order. Across the globe a substantial share of humanity continues to be afflicted by miseries such as poverty, warfare, homelessness, hunger, dysfunctional or corrupt governance, disease, lack of clean water, and waves of contagious illness. While the most virulent effects of the global Covid-19 pandemic have been waning, alarming new prospects of antibiotic-resistant disease have come into view. Even America, a comparatively affluent nation, groans under the burden of never-ending racist assaults and indignities, a plethora of mass shootings, and a post-truth, highly fractured social-political order wherein trust, honesty, and commitment to the common good are in short supply.

Among the many forms of environmental assault impinging on the planet, the climate change emergency is especially troubling. That's not only because the scale of this problem is so momentous, and because the cascading ills with which it threatens virtually all lifeforms on earth are so compounded. Not only because it challenges many of the prevailing financial, political, and industrial interests of our society. It's also because *reversing* the course of climate change over the next few years has by now become all but impossible. The excess of carbon dioxide and methane already assimilated into earth's atmosphere is expected to remain there for hundreds of years, even if we succeed in drastically reducing new emissions. According to the United Nations Intergovernmental Panel on Climate Change, we have at this writing only a decade or so left to us, if we act swiftly and decisively as a global community, to avert catastrophic results. The Panel concludes that we are unlikely in this century to limit warming to 1.5 degrees Centigrade and will be hard pressed, even with serious mitigation measures, to limit it to 2 degrees Centigrade.[1]

Even if international initiatives to reduce greenhouse emission were shifted promptly into high gear—as seems unlikely, given the current deficit of political will and leadership—it would already be too late to reverse the overall slide toward atmospheric degradation, too late to restore the wondrous old earth of preindustrial recollection, still largely in view just a few decades ago. The radically accelerated pace of greenhouse gas absorption into the atmosphere during the last thirty years means that this momentous transformation took place during an interval when the world had already been alerted to the peril that excessive emissions posed to life as we know it. Never has the human proclivity toward denial been so disheartening and so dangerous. Without some currently unforeseen breakthrough in technology, apparently the best outcome we might expect now is an appreciative mitigation of those otherwise

---

1. Report of the Intergovernmental Panel on Climate Change, accessed June 20, 2023, https://www.ipcc.ch.

grave debasements of planetary life that now seem likely. Our main aspiration now must be to avert global catastrophe during this century—if, that is, we can as a global community commit to taking decisive action toward carbon reduction, combined with some adaptation, in the course of this decade.

Evidently, too, the least privileged members of the human family will continue to bear the worst consequences of humanity's dereliction in this regard. As activist Peterson Toscano points out, the ecojustice consequences of climate change are such that "we may all be on the same boat, but we are not all on the same deck."[2]

How, in the light of all this, might we conceivably find grounds for investing hope in the future? Where, amid this bad news, could we still find the "Good News" that Scripture claims God has declared to the world? And on what terms, if at all, might it be possible for us to renew that deep green faith so sorely needed now both by us and by the earth we inhabit?

A response to these tough questions has to begin, I think, with noticing that the core teachings of Jewish and Christian faith have never been inclined to equate hope with a naively sanguine view of the future. In this context, hope, which St. Paul ranks beside faith and love in his trio of core virtues, is not just temperamental optimism. It's an aspiration that takes to heart the realities of loss, grief, and the cross. Environmental consciousness in this era of the Anthropocene must encircle, rather than deny, that which Mary Evelyn Tucker has called the "tsunami of sadness"[3] that engulfs everyone cognizant of the immense losses already suffered or projected to arrive in the wake of climate change.

---

2. Peterson Toscano, "Not Our First Rodeo: Memory and Imagination Stir Up Hope," in *Rooted and Rising: Voices of Courage in a time of Climate Crisis*, ed. Leah D. Schade and Margaret Bullitt-Jonas (Lanham, Maryland: Rowman & Littlefield, 2019), 85.

3. Mary Evelyn Tucker, Foreword to *Rooted and Rising*, xiii.

The most relevant Scriptural teachings on the matter portray hope neither as an innate personality trait, a certain disposition of mind, nor as an attribute that individuals might look to generate and cultivate apart from other values. Understood instead as an outgrowth of love and faith, it concerns in biblical terms a radical trust, an abandonment of will to a divine presence beyond ourselves. An abandonment of egotism, ambition, and false prospects, but *not* of human responsibility and co-creative imagination. The far horizon of such hope lies within the eschatological—that is, ultimate—mystery of redemption. Within this mortal life, hope remains, like that which it aspires to achieve, a work-in-progress rather than a stable presumption. "For in hope we were saved," St. Paul claims, though that hope invariably takes the form of "what we do not see," so that we must "wait for it with patience" (Romans 8:24).

For Paul, as suggested by his succinct declaration of Christ's death and resurrection in 1 Corinthians 15, "hope" is better defined as a habit of being, something meant to be practiced, exercised, and proclaimed rather than a wish. It is thus conceived to be more nearly a verb, a muscular movement, rather than a noun linked to some inherent trait that we have. Hope, in short, might well be considered something we *do*. For within Paul's supreme triad of virtues, all of the aspirations named by hope are viewed as inseparable from faith and love. All future hopes for humankind and the restoration of Creation are rooted, in turn, in Christ's resurrection from the dead, a past event bound up with what is still to come.

From the standpoint of this elusive version of gospel "hope," all human initiatives to overcome the climate crisis could be regarded as impossible—and yet, paradoxically, imperative. As the arresting headline of one op-ed contribution in the *New York Times* puts it, "Stopping Climate Change Is Hopeless: Let's Do it." Once again, the hope in question requires an active exertion of will, not just a wish for something better. For Christians, such hope also rests upon a fundamental faith, though surely *not* a passive presumption, that "God's life-saving, justice-seeking love is stronger than all else." As Cynthia Mode-Lobeda points out, "Christians have heard the end of the story." Accordingly, they have cause to appreciate how "the promise given to the earth community is that abundant life will reign. So speaks the resurrection."[4] Back in the 1960s, theologian Jürgen Moltmann likewise set forth a "theology of hope" that "strains after the future," that seeks inspiration beyond the world as we know it but "does not suppress or skip the unpleasant realities."[5]

Faith-inspired hope can be extraordinarily persistent. It refuses to quit even when quitting seems by far the most sensible course. The career of William Wilberforce, a famed antislavery legislator in Britain during the late eighteenth and early nineteenth centuries, offers a moving illustration of this. Following a relatively dissolute life as a young adult, Wilberforce experienced in 1798 the "great change" of a religious conversion that led him to become an evangelical Anglican. He entered politics, serving as a longtime Member of Parliament committed to "the suppression of the slave trade." He also supported the cause of animal welfare and helped to found the pioneering Royal Society for the Prevention of Cruelty to Animals. Yet from the outset, his initiatives in the House of Commons to pass legislation ending the slave trade met with determined resistance. For almost two decades, beginning in 1789, he campaigned tirelessly to achieve this end. Again and again throughout this period he renewed the struggle—soliciting allies while introducing and defending with eloquence various bills he hoped could be

4. Op-ed piece by Auden Schendler and Andrew P. Jones published in the *New York Times* on 6 October 2018, cited by Jim Antal in *Rooted and Rising*, 123; Cynthia Moe-Lobeda, "Love Incarnate: Hope and Moral-Spiritual Power for Climate Justice," in *Ecotheology: A Christian Conversation*, ed. Kiara A. Jorgenson and Alan G. Padgett (Grand Rapids, Michigan: Eerdmans, 2020), 82.

5. Jürgen Moltmann, *Theology of Hope: On the Ground and the Implications of a Christian Eschatology* (New York: Harper & Row, 1965), 19.

approved to enact the needed reform. Year after year he endured the frustration of seeing all his efforts defeated or found action on the matter deferred in one way or another. But he persisted, meanwhile enduring acute pain from a stress-related illness (probably ulcerative colitis) until finally winning through to success in 1807. And it wasn't until 1833, after he had left Parliamentary service and just three days before his death, that slavery itself was abolished throughout the British Empire.[6]

To my mind Wilberforce's story demonstrates how only a more-than-worldly hope for the world, undergirded by a larger vision of things, may be sustainable throughout decades of rejection and discouragement. Such hope surely has more to do, at any rate, with determination than with merely wishing or desiring. Creatures from the animal realm, too, can teach us something about single-minded persistence. John Muir was much taken, for example, by the irrepressible vigor and endurance of the water ouzel, a diminutive bird of the Sierra who impressed Muir by flying fearlessly into mountain torrents, and by continuing to sing sweetly regardless of storm or seasons. In a similar vein, poet Emily Dickinson chose a bird's tenacious bearing as her figure of active resolve when she wrote of how "Hope is the thing with feathers / That perches in the soul / And sings the tune without the words /And never stops at all."[7]

At the time Dickinson could scarcely have predicted, of course, the threats to aviary populations and diversity that human incursions on their habitats would bring. Yet by our own day, journalist Elizabeth Kolbert and composer Christopher Tin have both been moved to reappropriate her verses in

---

6. Wilberforce also advanced the cause of curbing animal cruelty in Britain. His career figures among the cases I address in the previously referenced essay, "'Not a Tame Lion': Animal Compassion and the Ecotheology of Human imagination," 200.

7. John Muir, "The Water-Ouzel" chapter of *The Mountains of California*, in *John Muir: Nature Writings*, ed. William Cronon, 486–499; Emily Dickinson, *The Complete Poems of Emily Dickinson*, ed. Thomas H. Johnson (Boston: Little Brown & Company, 1960), 116.

precisely that light—and by way of imagining the *sort* of hope suited to weathering the grievous trials of a new Anthropocene era. In "The Thing with Feathers: *Homo sapiens*," the closing chapter of *The Sixth Extinction: An Unnatural History*," Kolbert develops a darkly oblique reference to Dickinson's poem. And Tin offers his choral setting of "Hope is the thing with feathers" as a featured component of his larger composition, titled *Lost Birds: An Extinction Elegy*.[8]

Hope is the thing with feathers.

8. Elizabeth Kolbert, *The Sixth Extinction: An Unnatural History* (New York: Henry Holt and Company, 2014), 259–269; Christopher Tin's choral setting, as performed by the British vocal ensemble Voces8, accessed June 22, 2023, can be sampled at https://www.youtube.com/watch?v=5q_rRKC4pZk&list=RD5q _rRKC4pZk&start_radio=1&rv=5q_rRKC4pZk&t=42.

But what might the present-day practice of earth-hearted hope come out looking like, especially given the apparent hopelessness of achieving full environmental restoration? Hopelessness can, after all, become a self-fulfilling stance, even amounting to another form of climate denialism if it ends in passive resignation or despair. By entertaining visions of a sustainable future, though, we might look together not only to express but also to enlarge our capacity for hope. One such pathway of promise appears in steps recently made across the globe toward discovering and implementing new forms of renewable energy. Other encouraging steps include the efforts underway to develop regenerative modes of agriculture, a rewilding of ecosystems and ruined landscapes, and the restorative designation of oceanic "no fish zones."

The current crisis also encourages us to form individual patterns of life suited to the practice of earth-hearted hope. It's fair to ask what shape these personal practices should take. For the moment, I'll name just three major ways one might think to respond. One might, for example, look to intensify one's commitment to pursuing earth-friendly—thus hopeful and healthful—patterns of daily living for oneself. In the course of that pursuit, it helps to remain alert to the sundry, ever-changing practical questions that living on the earth today must raise for us about matters such as our relation to energy and land use, housing and transportation, toxic consequences of our domestic habits, and our inevitable involvement in food acquisition and consumption. Many print and online sources are available to offer guidance on these topics.

Becoming engaged in some sort of activist enterprise to advance environmental remediation—or, at least, voting deliberately and offering one's financial support toward these efforts—is another important expression of hope. Opportunities for civic involvement in this vein abound, whether in the local community or in broader spheres of influence. Here are a few publications one might consult for more detailed information about earthcare activism than my treatment here allows:

Gregerson, Joan. *Climate Action Challenge: A Proven Plan for Launching Your Eco- Initiative in 90 Days.* (Positive Energy Works, 2023).

Jorovics, Stephen A. *Hospitable Planet: Faith, Action, and Climate Change.* (Morehouse Publishing, 2016), especially pp. 127–150.

Schade, Leah D. and Bullitt-Jonas, Margaret. *Rooted and Rising: Voices of Courage in a Time of Climate Crisis.* (Rowman and Littlefield, 2019).

A third important option would be to focus on incorporating practices of earth-engaged personal meditation, Rule of Life, and sacramental worship more deliberately into one's ongoing spiritual formation. I address this expression of hopeful living in the following chapter. Moreover, the handful of options identified here are not, of course, mutually exclusive.

Interpreting the Biblical Charge to Exercise "Dominion"

Sustaining an ecological hope consistent with Christian faith requires constant discernment, including a willingness to wrestle with problematic features of our Scriptural heritage. One troubling phase of the biblical narrative requires us to confront the way in which Israel's founding involved the forcible, often violent conquest of Canaan and displacement of its inhabitants. The opening chapters of the Book of Genesis, too, have long raised flash points of controversy about the nature and course of creation over time, about how the scriptural narrative relates to scientific and historical accounts, and about how God intended us to live in fellowship with other creatures.

It may no longer be so difficult to appreciate how the biblical story of life's development in successive stages can be reconciled with the findings of evolutionary science. That time and space, with all the rest of creation, did not always exist but rather came into being is likewise a point of biblical teaching consonant with modernity's prevailing "Big Bang" theory of cosmogenesis. Apparently there was a real "beginning" to things. But some readers have been especially troubled by what they take to be the obstacle to green faith and ethics posed by the directive, in Genesis 1:26–28, for humans to "have dominion over" over the rest of creaturely existence and to "subdue it." How should we interpret these injunctions today?

As we have already noted, the biblical vision of "creation" concerns an ongoing process, one that is still underway and poised toward an ultimate fruition. Creation does not have to do solely, or even primarily, with the origin of things. To focus too exclusively on what tradition presents to us as Scripture's opening testimony in the Book of Genesis—or, for that matter, on just its first two or three chapters—risks distorting our larger impression of biblical teaching about essential things. While the Book of Genesis qualifies as a richly evocative, poetic, and storied text, offering us much to ponder, its teachings need to be understood in the much broader context of biblical teaching as a whole—and with reference to the discrete circumstances of time, place, and literary genre relevant to its composition.

How, though, might we find any hope or good news for earth in that specific mandate for us to exercise "dominion" (Hebrew *radah*) over other creatures? For admittedly, this directive has been invoked throughout history to justify all sorts of exploitative and destructive practices, as though offering our species lordly license to plunder the natural order however we might wish. So how might these oft-cited lines about "dominion" and "subduing" be more faithfully construed?

For one thing, it helps to distinguish rather sharply between the sense of "dominion," in the usual English rendering, and "domination." It has been suggested, in fact, that the common "dominion over" phrasing is better translated as "mastery among" the creatures.[9] Scriptural teaching as a whole underscores the point that the "earth is the Lord's" (Ps. 24:1), thus never to be regarded as the permanent possession of any human person or group. We are to understand the land given to Israel, or by extension to us, as conferred only provisionally and held in sacred trust. It makes sense with reference to a primal phase of humanity's struggle for survival, as it certainly does not for us today, for ancient biblical authors to be encouraging population growth and countering threats from the wild to human life and health.

But the key affirmation of green faith in chapter 1 of Genesis lies in its reiterated assurance: "God saw that it was good" (Gen. 1:9). This simple refrain is sounded three times before reaching its crescendo—"and indeed it was very good"—upon the full flowering of zoological diversity. What the refrain clearly brings home to us is the *goodness and godliness* of creation. But there is still more to be discerned about this blessing. Above all, it confirms that the manifold beings of God's own creation have value in their own right, not merely for the sake of their perceived value to human beings. They have, in the language of latter-day environmental ethics, not a purely instrumental or practical worth but an intrinsic worth, in and for themselves, and in God's eyes. Therefore, God "saw that it was good" for darkness, light, seas, and dry land to exist well before (indeed, we might say, billions of years before) the emergence of humans and in some real measure *apart* from that development. Light, for example, does not exist simply

---

9. Ellen F. Davis, *Scripture, Culture, and Agriculture: An Agrarian Reading of the Bible* (New York and Cambridge: Cambridge University Press, 2009), 55.

because our species happens to want and need illumination for the conduct of its affairs.

So the vision set forth in these opening chapters is appreciably more theocentric, or God-centered, than anthropocentric. It is telling that, within Genesis 1's mythologized sequence of "days" required for creation to unfold, humankind is not accorded a separate day of emergence but instead shares the sixth day with most other forms of wild and domestic animality. The likely older account, represented in Genesis 2, even suggests that human emergence preceded that of the animals. The first humans, formed of dust, are conspicuously portrayed there as outgrowths of the earth, as indicated by the linguistic interplay between the primal human's identification as *adam* and the *adamah*—that is, soil—from which he sprang. Moreover, the first covenant, said in Genesis 6–9 to be enacted by God through the medium of Noah and his family, is not with humankind alone but with "every living creature that is with you" and that shares the divine breath of life. It is fitting, too, that this pledge of solidarity with the community of creation should have been emblematized by a striking feature of nature: the rainbow. No wonder poet William Wordsworth rhapsodized at how "My heart leaps up when I behold / A rainbow in the sky," such that he could wish his "days to be / Bound each to each by natural piety."[10]

Yet the "dominion" language of Genesis 1 undeniably attributes a special role of oversight within the earthly order to our own animal kind and kin. That pivotal office includes naming—in effect, drawing to shared consciousness—other creatures, as well as tending and keeping (Gen. 2:15) whatever plots of living and growing things are entrusted to our care. "Dominion" involves a form of rule but one consonant with

---

10. William Wordsworth, "My Heart Leaps Up," in *William Wordsworth: Selected Poems and Prefaces*, ed. Jack Stillinger (Boston: Houghton Mifflin. 1965), 160.

the Old Testament presumption that human rule, likewise applicable to kings of Israel, should "reflect the qualities of God's rule, especially justice and compassion." The special role accorded to humankind within the conduct of earthly creation is not, therefore "one that lifts humanity out of the complex web of interrelationships that constitutes that community."[11]

From the first, the scope and character of this biblical commission of oversight was also qualified in several important respects. Originally, for example, the presumption in Genesis 1 had been that humans were to eat only plant-based food, not animal flesh. And particularly for Priestly authors of the creation account in that opening chapter, another point of restriction was crucial. Observance of the Sabbath, modelled on God's seventh-day "rest," presented a formidable, sacred limit to humanity's otherwise unlimited ambition to work and amass wealth, to achieve, to strive for more and more rather than resting in what had been given. Those persistent human ambitions and obligations, acceptable within bounds, could be allowed play throughout most of the week. But the Sabbath day was instead to be honored for the sole purpose of celebrating Israel's faith and the gift of God's creation.

In biblical perspective, what the right exercise of earthkeeping demands of human beings is cognizance of the essential responsibility they bear, surpassing concern for their own welfare, to other creatures and to the Lord of Creation. They alone among earthly creatures bear such responsibility, precisely because the environmental consequences of their behavior, for good or ill, are so momentous—as is all too evident today. What earthkeeping further demands of humans is a willingness to honor limitations. That includes limits to one's personal ambition and greed, both personal and collective limits to consumption or acquisition, and fundamental limits to

---

11. Richard Bauckam, "Being Human in the Community of Creation," in *Ecotheology: A Christian Conversation*, 28–29.

"growth" in the sphere of economic practice. Of course the planet's carrying capacity is itself limited, though humans have long resisted acknowledging that fact. They still do.

The upshot of the Eden story in subsequent chapters of Genesis dramatizes how drastically our first parents failed to practice the restraint and acceptance of limitation God expected of them as responsible earthkeepers. Planted in a domain of bliss full of life and beauty, they were granted access to the fruit borne from many trees from which they might choose. What an extraordinary privilege! They could eat freely from any of these trees—with just the *one* exception, pointed out to them. We can understand this resistance to accepting limitation as the main cause of their expulsion from Paradise. As Pope Francis observed in his *Laudato Si'* encyclical, "The harmony between the Creator, humanity and creation as a whole was disrupted by our presuming to take the place of God and refusing to acknowledge our creaturely limitations."[12] But as the larger course of history demonstrates, those mythic primal figures, Adam and Eve, are scarcely the only humans to have succumbed to, and suffered the consequences of, this original sin.

To recognize, finally, that our species bears a special role in determining the fate of earth might seem inconsistent with the vision of all we share in creaturely fellowship with those whom John Muir called our "earth-born companions and our fellow mortals."[13] This responsibility for oversight does not derive, however, from any monopoly of virtue or good sense on our part. Neither does it presume that creation exists for the sake of humankind, or deny the crucial role played by all other creatures within the ecological, geophysical, and astronomical ordering of the cosmos. Still the fact of the matter remains:

---

12. Pope Francis, *Laudato Si': On Care for Our Common Home* (Huntington, Indiana: Our Sunday Visitor, 2015), 48.

13. Muir, *Nature Writings*, ed. William Cronon, 826.

earth's future depends largely on us. Like it or not, that much has already been determined by evolutionary history. And while we cannot expect technology to save us, we must for the sake of a flourishing planetary future make thoughtful use of all that scientific learning and technical expertise makes possible. As theologian Ellen Davis points out, "The extent of our interference obligates humans to exercise what they [ecologists] call 'active management'—in the language of this biblical poem, 'mastery among' the creatures."[14] To suppose otherwise is to risk pursuing some sort of neo-romantic primitivism that simply cannot be sustained on an earth populated as it is today.

Beyond Stewardship: Finding the Right Language for Earthcare

Following from what has already been said about our place as members of God's community of creation, what language should be used to describe our discrete vocation within earth's household of life? What names for that vocation also enlarge the prospect of hope for the world, as we've been considering throughout this chapter? Language does matter. The terminology we deploy in connection with green faith not only serves to express that faith but also helps to define it. It's true that no words available to us qualify as totally adequate to capture the reality of these matters—or, for that matter, the ultimate reality of God's presence. Yet it's worth asking what particular assets and liabilities attach themselves to diverse ecospiritual terms we invoke along the way.

Within Christian faith-communities, "stewardship" language has become a favored idiom for expressing humanity's call to live out the biblical recognition that this "earth is the Lord's." The stewardship metaphor, though not my own

---

14. Ellen F. Davis, *Scripture, Culture, and Agriculture*, 55.

preferred option, has its virtues. It has some illustrative precedent in biblical parables and sayings, for example. It suggests that we're called to draw from earth's assets modestly and judiciously rather than arrogantly and with overbearing license. A good steward is conservative about expenditures and investments, liberal so far as possible in serving the needs and preferences of others. Above all, "stewardship" underscores the point that none of us owns that which we draw upon from the natural order. We are instead just proxy landlords of creation, charged only with *managing* its harvesting or extraction, its growth or preparation, and finally its use.

Yet this familiar stewardship language carries certain liabilities as well. There's sound reason, I think, to avoid relying on it alone to name our godly place on the planet. True, some stewards ancient and modern oversee people as well as things. But to regard ourselves primarily as nature's stewards can at least risk portraying the nonhuman world as a storehouse of lifeless substances to be judiciously dispensed. It risks obscuring our sense of that other world as community rather than commodity. It risks turning green ethics and policy into something of a resource-management project rather than an integrally conceived pattern of life inspired by wonder, love, and praise. The less salutary connotations of resource-management call to mind, in turn, the way Gary Snyder, in a signature poem of his from the 1970s titled "Mother Earth: Her Whales," satirizes the sort of utilitarian conservation ethos in which "robots argue how to parcel out our Mother Earth / To last a little longer."[15]

True, a good steward does more than parcel out material provisions. A good steward may manage personnel, too, showing respect and hospitality toward other persons as circumstances warrant. But stewards, though often highly trusted, are in essence employees. The oversight they are ordinarily expected

---

15. Gary Snyder, "Mother Earth: Her Whales," in *Turtle Island* (New York: New Directions, 1974), 47–49.

to bring to tasks they perform cannot match the heartfelt personal devotion that biblical stories show self-employed shepherds bringing to the care of their sheep. It cannot match, either, the sort of deep involvement and connectivity that human parents bring to oversight of their own family members.

In this respect two other terms commonly invoked today, "creation care" and "earthkeeping," strike me as preferable to "stewardship" when it comes to promoting the practice of green faith. "Creation care" benefits from the apt, twofold connotation that this term projects. It suggests, on one level, that the effective practice of green faith requires some form of attentive, diligent, concrete action, in much the same way we might "care for" another human or nonhuman being. In addition, it suggests how we're often invested emotionally in these labors and the living beings thus engaged, in the sense of "caring about" them.

An "earthkeeping" frame of reference offers multiple benefits as well. For one thing, such language is aptly grounded in the Genesis story, where the first man's clear assignment in Eden is "to till it and keep it." To "keep" God's creation, rather than to re-invent or supplant it, also defines the charge of human dominion in relatively modest terms, as constituting most essentially a task of *preservation*.

In addition to reflecting on these broadly encompassing titles for our practice of green faith, I believe it's also useful to identify a few turns of phrase evoking key attributes of that faith. I want to single out here just four of these idioms, most of which receive fuller treatment elsewhere in this book and all of which have been invoked in texts by living writers of note, including Wendell Berry and Pope Francis:

- our sacred *membership* in the great economy and community of life
- the *giftedness* of land

- our *sacramental participation* in creation's ongoing life, death, and renewal
- *reverence* for the earth

As indicated in the opening chapter, one figurative expression useful in defining our divinely willed place in the natural order concerns our membership in the great household of creation. Wendell Berry's writing often addresses this theme, underscoring the inherent giftedness of land. He expounds most directly on the divinely given—rather than proprietary or commodity-determined—character of Creation in his early, biblically conversant essay on "The Gift of Good Land." Like Aldo Leopold before him, Berry understands "land" in this context to refer not merely to a plot of soil, a purely material thing but to Leopold's expansive "revolving fund of life"[16] and energy, amounting, we'd now say, to an ecosystem. As a gift conditionally conferred, land offers humans the benefit of a sustainable, flourishing livelihood but likewise calls them to fulfill certain responsibilities. As beneficiaries of the gift, we're to understand our own authority and role in creation as radically contingent rather than absolute. Precisely because we hold such influence over earth's environment, ours is a sacred vocation.

Not only sacred but charged with sacramental meaning—that is, a relation set forth in classically Christian language as an engagement with nature's outward and materially visible signs of inward and spiritual grace. Henry Thoreau, too, like some other non-Christian but devout spirit seekers, commonly used sacramental language to express his ongoing communion with features of the natural world.

The way Wendell Berry articulates our sacramental engagement with life-and-death processes of the natural order in "The Gift of Good Land" strikes me as especially

---

16. Aldo Leopold, *A Sand County Almanac and Sketches Here and There*, 216.

penetrating. It is a statement aptly grounded in the Eucharist, a pivotal sacrament of Christian practice. Recalling first how "we depend upon other creatures and survive by their deaths," Berry points out that

> To live, we must daily break the body and shed the blood of Creation. When we do this knowingly, lovingly, skillfully, reverently, it is a sacrament. When we do it ignorantly, greedily, clumsily, destructively, it is a desecration. In such desecration we condemn ourselves to spiritual and moral loneliness, and others to want.[17]

For Pope Francis, too, the "sacramental signs" evidenced throughout the material world witness to how "the universe unfolds in God, who fills it completely." The Church's officially designated sacraments likewise enact the "way in which nature is taken up by God to become a means of mediating supernatural life."[18]

That the natural world deserves not only to be honored and preserved but also to be *reverenced* is another important corollary of deep green faith. In general terms, we practice *reverence* when we pay homage to the sacred worth that is discernible in something or someone beyond ourselves. The numinous presence we discern in that which we reverence confronts us with mysteries surpassing our understanding and control. It takes us momentarily out of ourselves. To that extent "reverence" bears a religious meaning that exceeds the ordinary civil virtue of *respect* we owe to all persons and many things. According to Christian belief, we reverence whatever warrants more than our respect but less than our worship. Roman Catholic theology, especially, commonly distinguishes between the

---

17. Wendell Berry, "The Gift of Good Land," in *The Gift of Good Land: Further Essays Cultural and Agricultural* (San Francisco: North Point Press, 1981), 281.

18. Pope Francis, *Laudato Si'*, 150–151.

worship or adoration (Latin *latria*) owed to God alone and the veneration or reverence (Latin *dulia*) properly shown toward saints and holy objects. We may therefore bow before the cross, an altar, and holy icons to express reverence, but in doing so intend to show worship only for the One they represent.

Around the middle of the last century, Christian humanitarian Albert Schweitzer set forth his conviction that the modern world needed to cultivate a "universal ethic of reverence for life." Although the boundaries of ethical concern and responsibility had rarely at this point in history been extended beyond human affairs, Schweitzer proposed a philosophic ethic applicable to "all living beings." "Who among us," he asked, "knows what significance any other kind of life has in itself, as a part of the universe?" For "To the person who is truly ethical all life is sacred, including that which from the human point of view seems lower." Schweitzer pointed out how "In the parable of Jesus, the shepherd saves not merely the soul of the lost sheep but the whole animal. The stronger the reverence for natural life, the stronger also that for spiritual life."[19]

Such declarations had a major influence on proponents of the modern environmental movement. Rachel Carson, for example, dedicated *Silent Spring*, her landmark ecological pronouncement deploring thoughtless pesticide use, to Albert Schweitzer.

Sometimes reverence for the earth and certain of its creatures can also be expressed outwardly through ritual gestures. Physical movements geared to the Sun Salutation, a sequence familiar to practitioners of Hatha Yoga, represent one such expression. Another is the gesture, expressing reverential homage to the material and spirit world one inhabits, of turning round to face each of the four cardinal directions,

---

19. Albert Schweitzer, "Out of My Life and Thought," in *Reverence for Life: The Ethics of Albert Schweitzer for the Twenty-First Century*, ed. Marvin Meyer and Kurt Bergel (Syracuse: Syracuse University Press, 2012), 117–118.

as described by Black Elk, the famed Lakota Sioux medicine man.

I especially appreciate, and have sometimes been inclined to emulate, how the writer-adventurer Barry Lopez reports his having embodied the spirit of reverence toward animals and landforms he encountered during his travels in far-flung Arctic settings. He first found himself bowing before a single horned lark, then to other birds he saw exposed on the tundra during his evening walks. Eventually Lopez practiced this ritual habitually, self-consciously: "I would bow slightly with my hands in my pockets, toward the birds and the evidence of life in their nests—because of their fecundity, unexpected in this remote region, and because of the serene arctic light that came down over the land like breath, like breathing."[20] His travel narrative, *Arctic Dreams*, concludes with his offering a grand bow toward the icy waters of the Bering Strait.

Within the context of Christian liturgy, this sense of reverence for the earth is perhaps most clearly enacted in features of the Great Vigil of Easter. The lighting of the new fire, the blessing of the paschal candle (including, in the ancient *Exsultet* hymn, praise and thanks to the bees who contributed its wax), the progression from darkness to light, the sanctification of baptismal water—all of this bespeaks a reverence for elemental facts of creation conjoined with worship of its Creator. In liturgical churches the sanctuary's characteristic east-facing architectural design—toward rising both of the sun and the Son—is another telling detail. So, too, is phrasing found occasionally in liturgical texts, including The Episcopal Church's "Prayers of the People, Form IV," beginning with the supplication to "Give us all a reverence for the earth as your own creation."[21]

---

20. Barry Lopez, *Arctic Dreams: Imagination and Desire in a Northern Landscape* (New York: Scribner's, 1986), xx.

21. *Book of Common Prayer*, 388.

Toward a Recovery of Sabbath Rest

What the Genesis story sets forth as the first fruition of hope and the culminating stage of God's first creation is the Sabbath rest achieved on the seventh day. This hallowing of the Sabbath has deep significance for Jewish and Christian faith, though present-day Christians often struggle to grasp that significance and its relevance to living in harmony with the earth. Even for those who are reasonably observant Christians, the day of Sunday Sabbath is apt to mean little more than a passing break from the usual work week. For many today it calls to mind dour restrictions rather than celebration, including outdated or extinct legal prohibitions against retail commerce and labor. Or else it is narrowly equated with just that hour or so spent in church worship. For Christians, Sunday, honored as the Lord's Day and eighth day of creation by virtue of Christ's resurrection, has always been seen as bearing a problematic relation to the Jewish *shabat* that it never quite supplanted. Yet the need to honor sabbath time as blessed remains fully applicable to Christian as well as Jewish communities of faith.

Moreover, the promised benefits of Sabbath-keeping extend well beyond humanity. Leviticus 25 mandated the observance in Israel, every seventh or Sabbath year, of a "complete rest for the land" when cultivation was to be suspended. Torah provisions for celebrating the hallowed Jubilee season of release and liberation, every fiftieth year, likewise confirmed that the land and all its creatures belonged ultimately to God.

Rabbi Abraham Joshua Heschel pointed out how, from the standpoint of faith, the Sabbath day signifies nothing less than the wholeness and fulfillment of Creation. Its justification is not mainly utilitarian, to provide an interval of respite to ensure a productive work week. As Heschel writes, "the Sabbath as a day of rest, as a day of abstaining from toil, is not for the purpose of recovering one's lost strength and becoming fit for the forthcoming labor." Rather, "the Sabbath is a day for the sake of life," an occasion when "we are called to share in what is eternal in time, to turn from the results of creation to the mystery of creation."[22]

As Heschel discerned, Sabbath time thus enacts the sacralization of time itself, since it is meant to be a period set aside from our normal schedule of labor, obligation, and anxiety. It is a time to let go of worries about the future, about what we are doing or meant to do or becoming. It is a time simply to be, to contemplate the joyous gift of Creation. So we have things backward if we suppose Sabbath time exists to make us better workers. It's rather that we work toward the end of achieving fullness of life under grace, for ourselves and ideally for all other beings as well.

Sabbath time exists solely for the purpose of celebrating and experiencing something of that fullness. Norman Wirzba beautifully encapsulates these points in declaring how "God's *shabbat* completes the creation of the universe—by demonstrating that the proper response to the gifts of life is celebration and delight."[23] And for Heschel, this sense of completion and fulfillment on the seventh day, identified in Jewish tradition as Menuha, is the essence of God's Sabbath gift to the world.[24]

It's nonetheless true that most of us, myself included, do not ordinarily find it easy, or even possible, to set aside all competing obligations for one full day each week. Our personal circumstances, combined with all the secular expectations pressing upon us, often squelch that lofty intent. Although we long for satisfying rest amid our many and anxious pursuits, we often find it hard to justify claiming it. Yet as Wirzba wisely grants, our commitment to honoring sabbath time "does not

---

22. Abraham Joshua Heschel, *The Sabbath: Its Meaning for Modern Man* (New York: Farrar, Strauss & Giroux, 1951), 14, 10.

23. Norman Wirzba, *Living the Sabbath: Discovering the Rhythms of Rest and Delight* (Grand Rapids, Michigan: Brazos, 2006), 13.

24. Heschel, *The Sabbath*, 22.

depend on the cultural sanction of complete rest for one day of the week."[25] Sabbath time can best be understood, in fact, as a way of life not to be confined within that seventh day. In his searching, wide-ranging commentary on what it might mean for us to live the Sabbath today, Wirzba goes on to explain how this outlook bears on many dimensions of our existence including work, our creative endeavors, recreation, festivals, worship, family relations, education, economics—and, of course, our relation to the natural environment.

Through our Jewish heritage, we already have faith-based reasons to appreciate how the land, too, deserves its own sabbath rest, especially through sound agricultural practices akin to crop rotation. The Hebrew Bible and teaching from the rabbis suggest to us as well some longstanding restraints against the human exploitation of animals. But the present-day conditions of life in a consumer society also impel those of us who are removed from participation in traditional agricultural rhythms to reflect more deliberately in a sabbath light on what we eat, where our food and water and energy come from, where our garbage goes, and how we might minimize the burden our daily habits impose on earth's other creatures. Living faithfully thus calls us not only to *take* sabbath rest for ourselves but also to *give it*, so far as possible, to other humans and nonhumans.

The "rest" at issue here is a state not of torpor but of fulfillment and achieved satisfaction.

To live in hope is already to anticipate this satisfaction, here and now, in contemplative gratitude for creation. Wendell Berry's array of "Sabbath Poems," composed mostly outdoors over the course of several decades, is one fruit of this contemplative spirit. In more modest terms, I also try occasionally to practice what I term an "electronic sabbath," removing

---

25. Wirzba, *Living the Sabbath*, 14.

myself for a time from the powered devices that otherwise absorb much of my attention. While working with undergraduates in a field-based course involving a fair measure of hiking and silent reflection, I also found students receptive to group practice of this discipline—though I had proposed to them not a full day's retreat from electronics but more like 10–20 minutes.

The contemplative blessing of "rest," as George Herbert writes of it in his poem "The Pulley," is at once the jewel of all God's blessings and a counterweight to our inherent restlessness. Such rest corresponds not only to the fulfilled destiny of each individual life but to the divine consummation of life itself. Herbert suggests that the dissatisfaction we're apt to feel with the remainder or "rest" of our experience—with the manifold activities, goods, and concerns that otherwise absorb our attention—may nonetheless draw us back, as though bound to a pulley, toward the only ultimate beatitude: rest in and with God. Herbert imagines God musing over this prospect, deciding on behalf of humankind to

> . . . let him keep the rest,
> But keep them with repining restlessness:
> Let him be rich and weary, that at least,
> If goodness lead him not, yet weariness
> May toss him to my breast.[26]

Sabbath time means to offer us a foretaste of this rest.

Let me admit, though, that living habitually into the observance of earth-grounded sabbath time takes more patience and determination than I can usually claim. Still, during at least one season of my life, while serving for a year as a Fulbright lecturer in Senegal, West Africa, I did chance to

---

26. George Herbert, "The Pulley," in *George Herbert: The Complete English Poems*, ed., John Tobin (New York: Penguin, 1991), 150.

experience the sublime peace and joy that *shabat* can engender. That land's cultural ethos all but forced my wife and me to slow down the usual pace of our activities, opening the way toward a contemplative oasis of calm.

Almost every Sunday there, in fact, my wife and I felt privileged to enter a Sabbath time unlike any other we had known. In the morning we joined some other residents of the city for Eucharist at the Martyrs of Uganda Church, in downtown Dakar. The service featured a spirited, musically accomplished ensemble of African choristers, with the words of celebration blended into a linguistic stew of French, the local Wolof language, and liturgical Latin. We then made our way onto a stretch of beach beside the Atlantic Ocean, where we shared space on the sand with another couple who had become our friends. Filling out the seaside picture were children frolicking on the beach, squawking gulls, and Senegalese adults fishing nearby or paddling their colorful pirogue vessels onto the sand. By afternoon we'd end up sitting for a good while, mostly in silence, gazing out at the ocean. That was all. But that, we found, was enough. It was all the sabbath repose we could have desired. After that, we shared an informal meal with our friends at a nearby Vietnamese eatery before heading home, tired from sun and water but inwardly rested.

Melville had his character Ishmael declare that "meditation and water are wedded for ever."[27] By virtue of the sea's presence before us and within us, week after week, we felt certain this was so. Oddly, too, it was through our involvement in this unfamiliar, economically depressed, and largely Islamic culture that we came to learn much about the grace of joyous gratitude for creation that marks Sabbath living.

---

27. Herman Melville, *Moby-Dick*, ed. Hershel Parker, 17.

Questions for Discussion and Reflection

1. What do you find most disheartening about the current state of the nation, the world at large, and the ecological health of our planet? What, if anything, nonetheless gives you hope?

2. What aspects of your Christian faith might you especially take to be "'Good News"—not only for yourself and those you love but also for the larger community of life on earth?

3. Have you ever felt inclined to display *reverence* for some nonhuman creature or landscape feature? If so, what in particular inspired your response?

4. What do you take to be the right character and degree of humankind's sway over other earthly beings and substances, over ecosystems, and over the planet as a whole? What language do you think best describes whatever form of human oversight you've identified as suitable?

5. Sabbath keeping in today's secular world poses a challenge not only for individuals but also for fellowships of faith and other communal bodies. How might communities with which you've been associated either help or hinder you toward experiencing a meaningful *shabat*, especially one reflecting gratitude for what Rabbi Heschel called the great "mystery of creation"?

CHAPTER 5

Green Perspectives on the Transfiguration of Christ and Practices of Personal Spirituality

The Transfiguration of Christ and Creation

In these two closing chapters, our focus of attention shifts from major points of doctrinal belief toward a look at some specific spiritual practices—moving, in other words, from eco-theology into ecospirituality. Yet by way of bridging that transition, I want first to highlight one last strand of theological teaching: the Transfiguration (Greek *metamorphōthe*) of Jesus. Celebrated as a notable feast day in the Church's liturgical calendar, particularly for Eastern Orthodox Christians, it derives from the gospel story of how Jesus was marvelously revealed as the Christ of God, and transfigured with divine light, after ascending a high peak in Galilee traditionally linked to Mount Tabor. Three of his disciples witness his appearing in the mystical company of Elijah and Moses, with whom he converses, while his clothing turns "dazzling white, such as no one on earth could bleach them" (Mark 9:3).

As related by all three synoptic evangelists (Matthew, Mark, and Luke), this evocative yet often overlooked gospel episode foreshadows the glory of Christ's Resurrection. But unlike the Resurrection's unseen transformation, it tells of an epiphany witnessed in real time by three favored disciples: Peter, James, and John. Through the eyes of these witnesses we, too, become figurative participants in the event, which is envisioned in Trinitarian terms. The biblical author of Second Peter testifies, in fact, that when the Father declared "This is my Son, my Beloved, with whom I am well pleased," at that very moment "We ourselves heard this voice come from heaven, while we were with him on the holy mountain" (2 Peter 1:17–18). A metamorphosis takes place here not in the identity of Jesus himself but in the way Peter, James, and John perceive the deep reality set before them. The mountaintop episode in question thus presents us with a model of transformative "seeing into the life of things"[1] quite relevant to our understanding of ecospiritual practice. For it, too, can stir an inward transformation for its practitioners.

At the first, most evident plane of meaning, the Transfiguration episode dramatizes the glorification and divine sonship of Jesus as the Christ of God. Jesus' central role in salvation history is reinforced insofar as he is pictured here at the apex of eternity—but as standing within time between Moses, sage of past prophecy, and Elijah, seer of a future fulfillment. Situated roughly at the midpoint of the gospel narratives, this poetically framed account also shows Jesus *trans*figured on the Mount just before leaving Galilee to be *dis*figured and killed in Jerusalem. So it embodies a glory bounded by affliction and the cross.

At a second plane of meaning, the episode has been taken to signify the elevation and transformation not only of Jesus, but also of human nature itself, under the promise of redemption. In this light, the late Archbishop Desmond Tutu wrote that, at a time of personal despondency amid the turmoil

---

1. Wordsworth, "Tintern Abbey," in *William Wordsworth: Selected Poems and Prefaces*, 109.

of South Africa's apartheid era, he once felt himself stirred to recover faith in "the power of transfiguration—of God's transformation—in our world." He points out that this realization came to him, in fact, on the Feast of the Transfiguration (August 6), as he sat in a priory garden surrounded by the otherwise bleak presence of browned-over grass in South Africa's winter season.

As he goes on to disclose, the Transfiguration mystery set before him at this time a transformative vision not only of the social order but also of the material world and the whole of creation. At what we might therefore call a third, still more encompassing plane of meaning, Tutu thus recognized how "the principle of transfiguration says nothing, no one and no situation, is 'untransfigurable,' that the whole of creation, nature, waits expectantly for its transfiguration . . . when it will not be just dry inert matter but will be translucent with divine glory."[2]

In the same vein, Michael Ramsey, former Archbishop of Canterbury, had earlier proposed that "A Gospel of Transfiguration," centered in the vision of Jesus transfigured on the mountain, "reveals that no part of created things, and no moment of created time lies outside the power of the Spirit, who is Lord, to change from glory into glory." We can thus understand the change Jesus' disciples witness in his clothing, a material substance now rendered luminescent, as indicative of what it might be like to apprehend a transfigured world.

Pierre Teilhard de Chardin adds another welcome voice to this chorus. "The Transfiguration has become a favorite feast of mine," he once wrote, "because it expresses exactly what I love most in Our Lord and what I expect most ardently from him. May the blessed metamorphosis of the whole creation also take

---

2. Desmond Tutu, *God Has a Dream: A Vision of Hope for Our Time* (New York: Image-Doubleday, 2004), 3.

place within us, and before our eyes."[3] It was apparently a particular feast day of the Transfiguration on August 6, 1923, in fact, that became the occasion for Père Teilhard's eloquent, earth-infused meditation, published under the title "Mass on the World."

We do not find this enlarged, ecospiritual, and even cosmic application of the Transfiguration story spelled out within the gospel narratives. Yet Christian tradition, over centuries of reflection, has effectively brought it to mind. And we find this visionary perspective bolstered by an array of theologians, imaginative writers, and iconographers. For some time now I have felt moved by the imaginative richness and spiritual force of the Transfiguration gospel, particularly through its artistic recreations in literature, music, and visual presentations.[4] One striking case of visual testimony to the Transfiguration can be found in the apse mosaic in the Basilica of St. Apollinaris in Classe, near Ravenna, Italy. Its vision of a renewed Creation transposes the mountainous setting of the Transfiguration event, as represented in the gospel narratives, to a verdant meadow scene. The landscape portrayed in this mosaic includes palm trees, flowered meadow terrain, birds, and other natural features characteristic of the actual Mediterranean region where the church is situated.

Greek and Russian icons from the Eastern Orthodox tradition offer another revealing window into the Transfiguration. In a well-known Novgorod icon from the late fifteenth century, for example, Jesus holds the central place within an ocular circle traditionally regarded as "the Great Eye of God."

---

3. Arthur Michael Ramsey, *The Glory of God and the Transfiguration of Christ* (London: Longmans, Green and Co., 1949), 147; Pierre Teilhard de Chardin, letter (my translation) written 8 August 1919, in *Genèse d'une pensée: Lettres 1914–1919*, ed. Alice Teillard-Chambon and Max-Henri Begoen (Paris: Bernard Grasset, 1961), 394.

4. I elaborate on these matters, along with others touching the Transfiguration theme, in *The Transfiguration of Christ and Creation* (Eugene, Oregon: Wipf & Stock, 2011).

The Transfiguration of Christ, 15th-C Russian Icon of the
Novgorod School (credit Alexander Blinov, Dreamstime.com)

He stands on a craggy peak of earth, positioned amid the one triptych portrait featuring himself, Moses, and Elijah; and another below imaging Peter, James, and John. He thus stands, in effect, at the horizontal crux of time and eternity. The iconographer effectually situates him, on the upper plane, between the Torah revelation of Moses from the past and the herald of Israel's consummated future. On the lower plane, three favored disciples strive to absorb the epiphany of the present moment. Portrayed meanwhile along the vertical trajectory of this sacred geometry, Jesus stands at a summit point between earth and heaven, on a fracture point of the *axis mundi* where he appears to mediate the energies flowing between God and the physical world.

Both within and beyond Christian tradition, mountaintop vistas have long been valued as offering a spiritually enlarged perspective on the earth we inhabit. That seems especially true following a demanding climb some distance to the summit, as I can confirm from my own hiking ventures into several of the Adirondack peaks and high country elsewhere. The Hebrew Bible abounds in mountaintop revelations.

For Eastern Orthodox theologians and Church leaders especially, the response of those witnessing Christ's Transfiguration manifests a freshly discerning way of *seeing the world*. The Transfiguration Gospel thus becomes something like a verbal icon, through which one catches glimmers of the promised New Creation. Thus for Bishop Kallistos Ware, "Within the Gospel story the Transfiguration of Christ stands out as the ecological event par excellence." Sergei Bulgakov, Dean of the Orthodox Theological Academy in Paris until his death in 1944, likewise makes a compelling case for heeding the revelation from Mount Tabor:

> What was it that the mountain, and the air, and the sky, and the earth, and the whole world, and Christ's disciples saw? What was the glory that shone round

about the apostles? It was a clear manifestation of the Holy Spirit resting upon Christ and in him transfiguring the creation. It was an anticipatory revelation of "a new heaven and a new earth—of the world transfigured and illumined by beauty. . . . It is good to be here—very good." This is how the world is created by the divine Providence, though it is not yet revealed to human contemplation. And yet on Mount Tabor it is revealed already. And this contemplation of the incorruptible, archetypal beauty, is the joy of joys, "the perfect joy." This is why the feast of Transfiguration is an anticipation of joy, the feast of beauty.[5]

How, then, might this "feast of beauty" become more fully incorporated into Church life, lending fresh inspiration to that renewal of green faith and spirituality so desperately needed in our time? One approach I'd think well worth pursuing would be to celebrate the Transfiguration more deliberately in parish liturgies as the major feast day it is, underscoring its promise as a special occasion for honoring God's Creation. Such a Day of Creation could be observed either on the Transfiguration's traditional calendar date of August 6, which I'd think preferable, or on the last Sunday of the Epiphany season, where it already holds a potential place in the lectionary schedules of the Episcopal Church but is not formally designated as a feast day.

The nearest thing our larger culture now observes by way of a widely recognized, collective festival of creation takes the form of a secular rather than Church-sponsored "liturgy" of sorts—namely, the annually designated Earth Day, on April 22.

---

5. Kallistos Ware, "Safeguarding the Creation for Future Generation," in *Transfiguring the World: Orthodox Patriarchs and Hierarchs Articulate a Theology of Creation*, ed. Frederick Krueger (Santa Rosa, CA: The Orthodox Fellowship of the Transfiguration, 2006), 89; Sergei Bulgakov, "The Exceeding Glory," in *A Bulgakov Anthology*, ed. James Pan and Nicholars Zernov (Philadelphia: Westminster Press, 1976), 191.

There is still room for Christian faith communities to follow suit. True, the custom of blessing animals around the time of St. Francis's feast day (October 4) is by now practiced in many churches. That is one promising development. Another is the celebration of Rogation Days, linked to agricultural customs, that takes place in some liturgical churches and features outdoor processions across the land. A third step forward, though it has yet to gain broader attention, is the recent ecumenical initiative toward recognizing a Season of Creation, commemorated during the period from September 1 to October 4. I believe there is nonetheless good reason for churches to begin observing one, genuinely celebratory equivalent of a Christian "Earth Day" that is anchored in a defining gospel text and builds upon time-honored tradition. The August 6 Feast of the Transfiguration, already designated in the Book of Common Prayer as a holy day of such significance that it takes precedence of a Sunday, seems ideally suited to assume that role.

Christianity's Legacy of Nature Meditation

A spirituality centered on cultivating some form of disciplined, personal engagement with the natural world has been practiced for centuries, both within and beyond Christian communities of faith. Such practices certainly predate our own age of ecological self-consciousness. The contemplative legacy they reflect has its vital counterparts in Jewish, Hindu, Buddhist, Native American, and other religious cultures. Celtic Christians of pre-industrial Europe, for example, without identifying any separate category of "nature," took for granted their kinship with the beasts, plants, geophysical and celestial bodies, and unseen spirits with whom they shared daily life in this world.

As we might expect, Christianity's legacy of earth-grounded meditation has important roots in Scripture. In the Hebrew Bible, for example, the Psalter reflects certain disciplines by which souls might enliven their faith and traverse an expected sequence of moods in converse with Israel's God. Such discipline included the cultivation of prayerful attentiveness to features of the natural or pastoral landscape. What came to be called "meditation on the creatures" in seventeenth-century England was already a familiar practice, then, for the Psalmist, who was moved with holy wonder to "consider" (KJV) the multitude of God's creatures spread across earth and sky. Viewed within this grand tapestry of Creation, humans must see themselves occupying only a modest though worthy place of honor:

> O Lord, our Sovereign,
> how majestic is your name in all the earth!
> . . . .
> When I look at your heavens, the
> work of your fingers,
> the moon and the stars that you
> have established;
> what are human beings that you are
> mindful of them,
> mortals that you care for them? (Ps. 8:1, 3–4)

Psalm 19 likewise begins with the author's having pondered with amazement how "the heavens are telling the glory of God, and the firmament proclaims his handiwork" in forms of speech surpassing human language. And throughout the early Christian and medieval periods, appreciation for earth-inspired forms of meditation grew, under the influence of monastic tradition and with the benefit of insights contributed by pivotal figures including Saints Augustine, Francis, and Bonaventure.[6]

---

6. I discuss this tradition more extensively in "Meditation on the Creatures: Ecoliterary Uses of an Ancient Tradition," in *Early Modern Ecostudies: From the Florentine Codex to Shakespeare*, ed. Thomas Hallock, Ivo Kamps, and Karen L. Raber (New York: Palgrave Macmillan, 2008), 181–192.

In today's popular usage, the word "meditation" can often mean just thinking about something, a mental activity akin to daydreaming. Understood as a spiritual practice, however, meditation becomes a decidedly *mindful* exercise. It aspires toward nothing less than the holistic integration of the self. That integration requires the interactive engagement of both mind and heart. It also commonly involves the self's mindful recognition of its rootedness within the larger, physical environment it inhabits. Meditation offers a technique by which we can hope to experience something of that larger reunification between ourselves and everything else that Emerson once defined as "the NOT ME."[7] It allows us to glimpse the restoration of a harmony God intends between human and nonhuman spheres of being, as well as between the Creator and Creation.

Meditation, then, is something that we *do*, quite deliberately. And while the words "meditation" and "contemplation" can sometimes be used interchangeably, I think it useful to distinguish one from the other. We might best identify "contemplation" not with something we *do* but with a state of our *being* or disposition. To develop a contemplative relation to ultimate reality—that is, to the world in God—means, in this sense, to progress toward becoming one with God. Few of us can expect to achieve such a union in this life.

We can learn much, though, from the accounts of meditative inquiry expressed in poems written by George Herbert and other astute devotional writers of seventeenth-century England. Meditation on the creatures, together with meditation on the self and meditation on the Scriptures, was a major focus of their prayerful pursuit. That form of nature contemplation is memorably represented in Herbert's poem "The Flower." On its face the lyric dramatizes its speaker's inward

---

7. Emerson, *Nature,* in *Ralph Waldo Emerson: Essays and Lectures,* 8.

progression from a season of depression, spiritual desiccation, and faltering belief to a springtime of regenerative life. But in coming to terms with that shift, Herbert pursues a meditative process steeped in botanical and seasonal imagery. "How fresh, O Lord, how sweet and clean / Are thy returns!" he exclaims, "ev'n as the flowers in spring."[8]

Such imagery well suits the temperament of this English parson who was particularly fond of gardening. "Who would have thought," he marvels, that despite his having suffered a long winter of desolation "my shrivelled heart / Could have recovered greenness"? Who would have thought that "now in age I bud again," and that "After so many deaths I live and write?" His musings thus lead him to understand how his own soul's regenerative course mirrors what is reflected seasonally in the natural order. Those flowering perennials he'd earlier loved had never really died, having instead for a season "gone / Quite underground," where they were graced to "see their mother-root."[9] So also, in effect, had he.

What marks this poem as a "meditation on the creatures" is the way its botanical imagery draws us beyond the usual sort of figurative comparison toward a more participatory engagement. This flower becomes, in other words, more than just another metaphor of the self. Herbert develops instead an almost visceral *identification* with that floral freshness he had come to experience through his own work in the soil. An identification, too, with the implacable fate of mortality and mutability that all humans share with plant and animal life. Having "recovered greenness," Herbert becomes effectually "green" himself, insofar as he recognizes greenness to be a natural outgrowth of grace and creativity. He had encountered the "mother-root" of all life. Significantly, he thus comes to locate

---

8. George Herbert, "The Flower," in *George Herbert: The Complete English Poems*, ed., John Tobin (New York: Penguin, 1991), 156–157.

9. Ibid.

his vision of ultimate redemption not in an ethereal "heaven" high in the skies but in a freshly flowering garden below, a paradise wherein he figures modestly but contentedly as just one more divine implant:

> There are thy wonders, Lord of love,
> To make us see we are but flowers that glide; [that is, pass silently away]
> Which when we once can find and prove,
> Thou hast a garden for us, where to hide.
> Who would be more,
> Swelling through store,
> Forfeit their Paradise by their pride.[10]

Though it demands some commitment to silence and solitude, the contemplative exercise of nature meditation should not be viewed as a self-absorbed, escapist preoccupation that sets us apart from action, reform initiatives, and right living in the world. It can enable us, in fact, to recover deeper rootedness in that larger, vital world of God's ongoing creation and redemption. For Christians it can also serve to heighten awareness of how, by virtue of Christ's incarnation, our planet's material richness of life forms and processes never ceases to embody the Source of all life.

As the Trappist monk Thomas Merton exemplified throughout his own vocational experience, the lives of contemplation and of action should be seen as complementary rather than contrary. For as Merton wisely observed back in 1971,

> He who attempts to act and do things for others or for the world without deepening his own self-understanding, freedom, integrity, and love, will not have anything to give others. . . . We have more power

---

10. Ibid.

at our disposal today than we have ever had, and yet we are more alienated and estranged from the inner ground of meaning and of love than we have ever been. The result of this is evident. We are living through the greatest crisis in the history of man; and this crisis is centered precisely in the country that has made a fetish out of action and has lost (or perhaps never had) the sense of contemplation. Far from being irrelevant, prayer, meditation, and contemplation are of the utmost importance in America today.[11]

Some fifty years later, as we find ourselves engulfed by an environmental crisis even more unsettling than much of what Merton lived to see, these words ring all the more true.

Notes Toward the Practice of Earth-Engaged Meditation

How, given this inspiring legacy, might one develop his or her own practice of earth-engaged meditation? And how should one begin?

It is worth experimenting at the outset with various forms of faith-inspired nature meditation. Yet because concentration and attentiveness are key features of any meditative practice, a common first step would simply have you sitting comfortably alone in silence, whether outdoors or within walls, for a brief interval of some 10–20 minutes. Sitting still is no easy thing for many of us, given our conditioning in today's frenetic, electronically possessed society, so it helps to keep your sessions brief at the outset. During this exercise you're striving first to quiet your mind, then to narrow your gaze and devote full attention toward some particular natural object or

---

11. Thomas Merton, *Contemplation in a World of Action* (New York: Doubleday, 1971), 165.

organism set before you—a rock, for example, or feather, nut, fern, flower, tree branch, a stream or larger body of water. The primary aim is to absorb as best you can, with mind and heart, the "thisness" of your focal object, with some further thought of its place within the larger expanse of divine Creation.

To help me adopt the right mood for this sort of meditative exercise, I find it helpful to recall how poet Walt Whitman, toward the start of "Song of Myself" in his *Leaves of Grass*, pictures himself as settled first into a state of pure and patient receptivity. He aspires to do virtually nothing except to "loafe" at his "ease observing a spear of summer grass" while attending to his own bodily vitality—"respiration and inspiration, the beating of my heart, the passing of blood and air through my lungs." To heed one's own breathing, heartbeat, and bodily impressions in this way is, in fact, a time-tested way of grounding the meditative project. For centuries now, guides from Hindu, Buddhist, and medieval Christian cultures have all been sharing their wisdom teaching about diverse techniques of contemplative breathing. Whitman's absorption with that single blade of grass, as demonstrated in the poem, eventually allows him to expand his interaction with the larger world toward cosmic awareness of how "a kelson of the creation is love," and how he is himself entwined with "the hand of God."[12]

I find that sitting to gaze on the water flow over a stream-nestled boulder works especially well for this exercise. The sight, sound, and cold spray of water coursing over rocks speaks to me of that which endures in creation apart from my existence, as well as that which is new, ever changing. In the presence of flowing water, I am reminded, too, of the many scriptural allusions to wells, streams, rivers, and springs that all witness, in one way or another, to the primal Fountain of life.

---

12. Whitman, "Song of Myself" in *Walt Whitman: Leaves of Grass*, ed. Jerome Loving (New York: Oxford University Press, 1998), 29–30, 32.

In addition to these sitting exercises, Christian tradition has also recognized the spiritual potential of strolling mindfully outdoors, in solitude and silence. This practice, similar to Buddhist walking meditation, is reflected as well in some forms of holy wandering, or pilgrimage journeying. Part of this practice's soul-sustaining value derives from the way those repetitive leg strides can help to restore our sense of nature's rhythms and of our own animal vigor.

Another form of nature meditation in motion relies more on chance discovery. To begin practicing it one has only to venture outdoors, look around across all available planes of vision, and try to see with full attention whatever is to be seen—particularly within the nonhuman realm. The main challenge of this exercise is to see and revere that particular face of creation, however modest and familiar, as though seen for the very first time. A first step, then, is to absorb inwardly the presence of whatever organic or inorganic features draw our awareness most pointedly during the outing. You might then look to situate that recognition, more imaginatively, within some dimension of our faith tradition and experience. I can recall, for example, how I felt during a forest walk one chilly New England winter, when I suddenly came across a stand of elegant Christmas ferns jutting out from banks of fresh snow. The scene held me for several minutes. I was struck by its splash of vernal color amid whiteness. And I was drawn to reflect not only upon the amazing resilience of these ferns, and their ties to primordial natural history, but also upon the apt association of their name with the mystery of Christ's incarnation.

For many persons, this mostly freewheeling approach to outdoor reflection can be usefully augmented by programs that engage participants in an extended sequence of themes–with attributes like thanksgiving, reverence, or humility—by and through which to orient their meditative experience in each session. Robert Gottfried and Frederick W. Krueger's *Living in an Icon* volume offers sound guidance for such a program.

Not every form of creation-conscious meditation needs to be conducted outdoors, however, or needs to be inspired by immediate sensory experience. Since creation lies all around and within us, encompassing the civilized world as well, even practices apparently remote from direct engagement with nature may contribute toward a robust creation spirituality. Saying the Church's Daily Office on a regular basis can, for example, share in that growth to the extent that it enhances one's involvement in the unfolding liturgical year, a calendar linked in turn to ancient agricultural festivals, nature's cyclic rhythms of seasonal change, and the alternation of day and night.

There is also much to recommend about meditative exercises, sometimes conducted indoors, in which the practitioner sits quietly for a set interval with eyes closed, likewise removed from most other forms of sensory stimuli. One such exercise of meditative prayer happens to be an integral feature of my own spiritual practice. I take it outdoors when weather permits. For me it includes the inward repetition of a brief, mantra-like verbal formula, together with an interposed pattern of mindful breathing. This verbal formula is based on the "Jesus Prayer," with its famed invocation of the Divine Name and venerable standing in Eastern Orthodox tradition: "Lord Jesus Christ, Son of God, have mercy on me a sinner."

But for use in the personal exercise at hand, I have been moved to substitute this radical distillation of the classic phrasing: "Lord Jesus, have mercy." The reasoning behind my adaptation should be fairly evident. What that elemental plea for divine mercy amounts to, after all, is a prayer of the heart we must ultimately be inspired to apply not solely to ourselves but to the larger human family, both living and dead—and indeed to the entire community of God's creation. "Lord Jesus, have mercy"—an invocation stark in phrasing, though multilayered in meaning. Its plea for mercy is one I like to envision rippling outward—from myself to persons and circumstances I know well, then to others currently in need, and finally toward non-human faces of creation that the occasion calls to mind.

There seems to be precedent and warrant, even from the standpoint of Eastern Orthodox guides, for using the very basic verbal formula I have been describing. The irreducible essence of the Jesus Prayer is, after all, simply a prayerful invocation of the divine name. Doing that in the right spirit is what matters. Bishop Kallistos Ware thus assured seekers that "The verbal formula can be shortened" since "the one essential and unvarying element is the inclusion of the divine name 'Jesus.'" Otherwise, "each is free to discover through personal experience the particular form of words which answers most closely to his needs."[13]

As I see it, the point of this exercise is not so much begging God to become yet more indulgent toward me, in the face of my failings and griefs. It has much more to do with my willingness to participate in God's own will to spread abroad, ever-more widely and deeply, the Creator Spirit's loving solicitude for all that God has made. That unadorned cry, "have mercy," also resembles the bids for compassion and aspirations toward the sustenance and welfare of all beings commonly expressed in Buddhist traditions. It therefore qualifies as one of the most encompassing yet quintessentially Christian petitions we could hope to present.

Why a Rule of Life?

"How then shall we live?" For good reason, this familiar question, often attributed to Leo Tolstoy, continues to resonate today. It poses a challenge not only to us collectively, as

---

13. Kallistos Ware, *The Power of the Name: The Jesus Prayer in Orthodox Spirituality* (Oxford: SLG Press, 1974), 5.

members of a given culture and nation, but to each of us individually: How then shall *I* live? How then shall I live if I know I must finally die? How then, if I have embraced the teachings of Christianity? How then, if I am called to live responsibly on today's earth in ecological peril?

Unless I allow whim and happenstance alone to determine where I am headed from here, these existential queries lead me to confront, in turn, the very practical question of *how* I will decide to order and govern the future course of my life. That's where the idea of creating a personal rule of life starts to sound plausible, maybe even appealing.

For many centuries now, some form of a rule of life has been adopted to shape communities defined by Jewish, Christian, and other faith traditions. A Community Rule has been found, for example, among the scroll writings left by members of the Qumran settlement from the period of Second Temple Judaism. The sixth-century *Rule of St. Benedict* has been particularly influential in identifying patterns of life adapted through the years by many Christian monastic communities.

Although the very mention of a "rule" stirs resistance from many today, the term in proper context connotes something distinct from rigid legalism. A rule in this sense has to do with formulating certain standards by which to define and implement our communal or individual priorities. Its aim is to help us live more intentionally, so as to embrace more fully for ourselves and others that abundance of life Jesus says he came to bring. A rule of life is not, in essence, therefore, a set of laws meant to restrict behavior but a sacred yet highly practical method of realizing our deepest desires and ideals.

Those disinclined to set themselves under a "rule" may prefer to imagine such a pattern of deliberate living as a "way," consistent with an array of biblical idioms urging us to discover our own blessed pathway through life leading toward fulfillment in God. In this light, it is also worth recalling how, according to the Acts of the Apostles, proponents of the very faith we profess had first been called followers of "the Way."

Drawing upon monastic precedents in fashioning a rule suitable for individuals requires, in any case, a fair measure of reconsideration and adaptation. A personal rule of life can take many different forms. Plainly, too, an individualized rule offers plenty of room for variation depending on particular needs and preferences. But in considering next the prospect of conceiving a green rule of life, I think it worth keeping in mind two principles relevant to formulating almost any form of personal rule. One is our common need to recognize, though outside the usual monastic setting, our essential relatedness to community—our membership, in fact, in manifold communities of life, both human and nonhuman. Another is the incorporation of certain traits likely to appear in the most enduring and useful specimens of a personal rule, despite many variations in form and provisions.

Conceiving a Green Rule of Life

What a few of us have lately begun calling a Green Rule of Life is simply a Rule that recognizes ways in which our hearts and minds might be more consciously engaged in living with, and promoting the welfare of, the whole of God's creation. Despite its coloring, a Green Rule is not something distinct from the broader plan for deliberate living we have already been describing. But a Green Rule does, more explicitly than some other plans, include provisions that address our role and obligations as members of the larger household, or *oikos*, of creation at large. Like any other personal Rule of Life, it is best formulated prayerfully, deliberately, and in light of the ways in which our own progress toward spiritual health and wholeness is tied to that of other beings.

Despite innumerable variations in format and content, versions of the Rule that promote more abundant life are also apt to share a few features. For example, they remain constantly open to revision. Like the ancient rule of St. Benedict, they reflect a spirit of moderation—rather than heroic asceticism—by virtue of which their provisions are more likely to be practiced over the long term. Persons living alone or in community who avoid consuming animal products might decide, for instance, to allow for an exception if invited to share someone else's hospitality at a meal gathering. Yet the Green Rule's provisions must be challenging enough to address issues bearing on one's personal failings, addictions, and patterns of inertia. These provisions should embody, too, a gospel-inspired simplicity of heart. Instead of spelling out every detail of intentional living, they list governing principles. They recognize the seeker's need for silence and solitude. They also suggest particular times or occasions when given practices would be undertaken, since the best of intentions often wither under the fantasy of doing them "sometime" or "later." And they reflect, with respect to their "green" orientation, not only the subject's membership in multiple communities of life, including the community of all creation, but also gratitude for our existence on this planet.

In developing a Green Rule that includes a listing of key items and provisions, a person might choose to adopt one of several possible schemes of organization. A model I take to be especially apt for the purpose centers on the theme of relatedness. It has to do with the ways in which we, as human creatures, relate not only to our Creator but to all manner of other creatures, substances, and material or immaterial goods.

Some readers might want to consider forming a Green Rule in accord with this more overtly relational model. Here, then, is a sample outline of how such a document might be configured:

Model of a Green Rule Grounded in Relatedness

This model centers on the idea of interrelatedness, an inherently ecological and theological principle. None of us exists in a state of isolated autonomy. How exactly, then, can we best live out our organic connection to God, to our own bodily and mental being, to our human neighbors across the earth, and to the larger community of God's Creation? How can we best discover and appreciate our place as members within the greater household of all created beings? Working with this model aims to help us respond to such questions.

1. Growing our ties to God and spiritual practice.
   e.g., with respect to prayer, meditation, group worship, spiritual reading.

2. Growing our ties to our own physical and mental health.
   e.g., with respect to exercise routines, wellness, opportunities for play, refreshment, and patterns of sustenance (including food choices, sleep, etc.).

3. Growing our ties to fellow human beings, both near and far.
   e.g., regarding our contributions to the common life of communities to which we belong by choice or by circumstance; our support for charitable organizations or sociopolitical enterprises across the globe; our actions in support of care for family, friends, associates, or strangers in need.

4. Growing our ties to the larger community of life, and to the earth we inhabit.
   e.g., with respect to our consumer choices; gardening and procuring food; our provisions for housing, home maintenance, travel, and transportation; our role in preserving, nurturing or restoring land; our care for domestic and other animals.

One benefit of intentionally ordering one's life in accord with a Green Rule is the enlarged opportunities it offers for appreciating how often behaviors conducive to the health of earth's ecosystems also contribute to one's own physical and spiritual well-being. Replacing, so far as possible, habits of local travel by personal motor vehicle with walking, bicycling, or public transport is a case in point. The exercise gained counts as beneficial in more ways than one. Much the same could be said about avoiding the consumption of red meat and highly processed foods.

It is worth reiterating, finally, that whatever Green Rule and meditative expressions of ecospirituality one adopts should be viewed as incorporating, rather than replacing, mainstream practices of the Christian "way." The traditional meditative exercise of *lectio divina*, for example, based on reading and then absorbing at length discrete biblical passages, can easily be conjoined with other practices more obviously attuned to an ecological consciousness.

Other sorts of reading, too, can contribute substantially to our knowledge and soul-sense of Creation, especially insofar as the very act of reading silently in solitude requires a fair measure of meditative discipline. Not only theological or devotional texts but also exposure to many kinds of literary, scientific, or naturalistic writing can expand our appreciation of the wonders that permeate this cosmos we inhabit. For me, the commentaries published by certain astrophysicists, biologists and geologists, some of whom I happen to know personally, together with writings by a whole host of naturalist authors and adventurers from the past, have all proved invaluable to expanding my vision of the wider world.

The ideal represented in many traditions of faith-inspired meditative practice, particularly in the West, is, after all, a holistic apprehension of the world. And part of that self-integrative process requires an exercise of mind, as well as of the heart and soul. It necessarily involves us in deepening our engagement both with the life of faith and with the full span of God's Creation.

Questions for Discussion and Reflection

1. How might the New Testament story of Jesus' Transfiguration on a high mountain in Galilee inspire our own aspiration to "see" the world in a freshly expansive light, both literally and figuratively?

2. One occasion in our common calendar meant to express gratitude and appreciation for the created order we inhabit is Earth Day, April 22. Plainly, though, this global feast day has a secular origin and character. What faith-inspired equivalents of an "Earth Day" observance might therefore strike you as most promising for the Christian Church to designate and make integral to its life of worship?

3. What forms of creation-grounded meditation have you already thought to practice thus far, or want to consider adopting?

4. George Herbert, in his poem "The Flower," goes beyond simply describing a particular organism to pursue a deeply personal, imaginative identification with it. Can you recall responding to any similar impulse in your own experience? If so, what creature(s) inspired that movement of sympathy, and what might you have learned from the experience?

5. Whether or not you have already adopted or plan to grow a Green Rule of Life for yourself, what benefits could you imagine flowing from this practice?

CHAPTER 6

The Eucharistic Feast of Creation

First Things: The Relevance of the Eucharist to Green Faith

As indicated earlier in this study, I believe that a religious faith worthy of the name must not only speak to one's personal condition and aspirations but must "encompass everything" in its vision. It must, in other words, have something to say about the Mystery we call God as well as the vast plenitude of God's creation throughout space and time—including all life, matter, energy, cultures, and ideas.

By the same token, the Church's primary sacraments address our most fundamental experiences, needs, and longings even as they unite us to God and to one another in bonds that can never be broken. Baptism serves as a rebirth into Christ, a participation in his death and resurrection, which defines and shapes our lives from that point on. It effects forgiveness of sin; creates a force field of spiritual connection with the communion of saints across time; and by making us members of Christ's body, situates us within the fellowship of the Church.

The Holy Eucharist, too, has expansive significance. It renews the transformative graces of baptism and continually reconstitutes the Church in her most profound identity as the Body of Christ. It draws together the Word received and communion shared, the salvation story and bodily action. It also embraces the whole of creation in its supreme act of "oblation" or offering, as we shall see. Fortunately, modern currents of liturgical renewal, flowing through many denominations, have recovered the early Church's recognition that the Holy Eucharist is "the principal act of Christian worship on the Lord's Day and other major Feasts," as stated in the *Book of Common Prayer*.[1]

Christians who are zealous advocates of earth care may nonetheless feel dissatisfied with the forms of worship traditionally practiced in mainline churches. They may with some reason regard these forms as too anthropocentric in texture, as evoking too little of the natural order and the cosmic expanse of God's creation. One response to this deficiency is simply to supplement or modify existing eucharistic liturgies to take account of whatever creative-evocative options for hymnody, intercessory prayer, or preaching might be available within the faith-community's tradition. Another common response is to construct wholly new forms of earth-grounded worship, often devised for the benefit of "green" gatherings and special occasions. These initiatives strike me as promising as far as they go, particularly when the devised liturgies are adopted for non-eucharistic occasions.

Spearheading the creation of experimental liturgies does have its drawbacks, however. Such forms of worship rarely draw to prayerful song worshippers who are not members of the self-selected "choir." They may therefore be perceived and experienced as more of a special interest exercise or rallying platform than an occasion of common prayer. Classic liturgy, after all typically evolves slowly and organically from a given cultural context. Not so often from committee brainstorming.

Similar constraints are likewise worth acknowledging when it comes to preaching. Enlivening, genuinely revealing

---

1. "Concerning the Service of the Church," in *The Book of Common Prayer*, 13.

homilies play a crucial role in congregational development. Yet it takes imagination and artful grace to help one's hearers grasp what the Green Gospel demands of them while awakening, at the same time, gratitude and love for Creation. Frontal assaults on a congregation's perceived indifference or skepticism are unlikely to persuade anyone of anything. They are more likely to provoke resistance or even backlash.

But what I believe matters most, regarding the Eucharist's relevance to creation care, is discerning the real but sometimes hidden dimensions of creation spirituality embedded in existing liturgies. What sort of instructional formation might allow communicants to become more deeply and mindfully engaged with those dimensions? What touchstones of earthy materiality and cosmically expansive meaning are already present in our traditional forms of eucharistic worship, though they may be rarely recognized as such?

Such are the questions I particularly want to address here. Our sense of familiarity with liturgical texts can often, it seems, block realization of how forcefully yet subtly they expound a sacramental vision of creation—including the advent of God's New Creation. Accordingly, I think it worth exploring in this chapter just how the eucharistic action, as enacted within existing worship forms, shares in the larger life of all Creation.

My own life experience persuades me that encountering this sort of liturgical commentary, blended with a measure of personal reflection, really can deepen one's worship experience. As a young person attending Masses celebrated in Latin during the last gasp of Roman Catholic practice prior to Vatican II, I relied on written commentary as well as on the translations found in my Missal to tell me what, beneath the surface, the eucharistic drama was all about. At least some ordinary Christians in medieval England derived similar benefit from guidance provided in *The Lay Folks Mass Book, or,*

The Manner of Hearing Mass: With Rubrics and Devotions for the People. In much the same vein, it seems important for us today to re-envision the ways in which "the central act of Christian worship"[2] embodies God's love and saving intent for the whole of God's Creation.

The relevance of the Eucharist to creation spirituality also deserves special notice because of latter-day ecumenical progress toward acknowledging its standing as the Church's central act of worship. Though the rite still bears diverse titles across denominational traditions—as the Mass, the Lord's Supper, Holy Communion, or the Divine Liturgy—a remarkable convergence of views about its essential character has been achieved in ecumenical reports such as the Faith and Order Commission's 1982 landmark statement from Lima on *Baptism, Eucharist and Ministry.* By now, at least at official levels of doctrinal representation, much of the partisan bitterness that once drove disputes about the nature of Christ's eucharistic "presence" has faded. There is general cross-denominational agreement that this "central act of the Church's worship" expresses "great thanksgiving to the Father for everything accomplished in creation, redemption and sanctification" and is "the great sacrifice of praise by which the Church speaks on behalf of the whole creation."[3] By common consent the Eucharistic action recalls—but also re-enacts, by way of *anamnesis,* or living memory—the communion of worshippers with the crucified yet Risen Lord.

Those who recall the ancient liberation of Hebrew forebears from Egypt in the course of celebrating Judaism's Passover Seder are urged to regard themselves as contemporary participants in that rescue. The principle of Eucharistic *anamnesis* likewise underscores the immediate personal involvement

---

2. *Book of Common Prayer,* 388.
3. *Baptism, Eucharist and Ministry,* Faith and Order Paper No. 111 (World Council of Churches, Geneva, 1982), 10.

of communicants in Christ's death, resurrection, and initiation of New Creation. The Eucharist thus traverses those divisions we ordinarily perceive across space and time, opening the way toward what theologian John Zizioulas has called a "remembrance of the future."[4] We can also deepen our engagement with this mystery wherein time touches eternity by living into seasonal cycles of the liturgical year, an orbit of communal experience rooted in ancient Hebrew agricultural festivals and practices.

How, though, might we find even in standard eucharistic celebrations fitting expression of a "green" faith crucial for our time? Admittedly, the core features of eucharistic liturgy may not strike us initially—apart from those earthy essentials of bread and wine—as particularly evocative of the natural world. In classic outline, the rite most evidently calls to mind humanity's role in sacred history rather than an impression of the physical world we inhabit together with other creatures. Yet by probing the larger import of familiar eucharistic texts, we can discern some revealing green dimensions of this sacramental mystery.

The vision of divine creation conveyed in these texts is broadly encompassing. Its reach is, in fact, cosmic—a perspective on Nature that's typically distanced rather than localized or expressly variegated. It reflects multiple registers of reality, things both "seen and unseen," imagined with possible reference to all times and places. Jesuit scientist-theologian Pierre Teilhard de Chardin captured this materially grounded yet cosmic, radically transformational essence of the sacrament in a prose meditation of his, "Mass on the World," composed in response to his time on expedition in Asia's Ordos desert in 1923. Lacking the usual liturgical elements, Teilhard as faithful priest found himself envisioning as his altar "the whole earth," and

---

4. John D. Zizioulas. *The Eucharistic Communion and the World*, ed. Luke Ben Tallon (New York and London: T&T Clark, 2011), 58.

accepting as his paten and chalice "the depths of a soul laid widely open to all the forces which . . . converge upon the Spirit at daybreak." Convinced of the transformational process that follows from this perspective, he marvels at how "the immense host which is the universe is made flesh," at how "through your own incarnation, my God, all matter is henceforth incarnate." He even writes of how all creation thus assumes for him "the lineaments of a body and a face—in you."[5]

As suggested by Père Teilhard's testimony, with its allusion to Christ's incarnation, the Eucharist might well be regarded not only as a designated sacrament of the Church, but also as *the* consummate sacrament of Creation. Moreover, a sacramental vision of the world, as set forth by many theologians in both Eastern and Western spheres of the Church,[6] calls us to recognize the ways in which not just formal sacraments but many outward objects, images, creatures, and actions offer us inward signs of grace.

Other cosmically aligned features of this eucharistic vision I want to underscore include:

- a biblically founded recognition that *Creation* encompasses not only the opening Genesis accounts, referencing that which the Creator first brought to being, but also the open-ended emergence of a *New Creation*, wrought especially through Christ and by his Resurrection. "Creation" in its fullest sense thus describes a continuous process, not just a one-time event linked to primal origins.

---

5. Pierre Teilhard de Chardin, *Hymn of the Universe*, trans. Gerald Vann (New York: Harper & Row, 1965, 1969), 19, 24–25.

6. David C. McDuffie, in *Nature's Sacrament: The Epic of Evolution and a Theology of Sacramental Ecology* (Washington, DC: Christian Alternative Books, 2020), comments on several of the Western theologians who have advanced this perspective in one way or another—including William Temple, John Macquarrie, Edward Schillebeeckx, and David Brown.

- a presumption that Christianity's best motives for Creation Care include but also surpass the dutiful moralism conventionally identified with "stewardship." Eucharistic celebration, as a communal expression of praise, thanksgiving, and sacrifice, should be chiefly inspired by love rather than duty. The core sense of "Eucharist," after all, is an act of joyous thanksgiving. And as an act of collective worship, its force derives more from a *doxological* (that is, an offering of praise) than a morally admonishing impulse.
- Eucharist is all about *communion*—most critically and obviously, enactment of the believer's communion with God in Christ. But this inherently ecological principle of interconnection likewise defines an ideal communion *among* worshippers, recalling by extension even our relatedness to every other member—human or otherwise—of God's created order.

In the light of these principles, I think it worth identifying some noteworthy "green" features of a traditional eucharistic liturgy, with reference mainly but not exclusively to current Episcopal texts throughout successive stages of the "ordinary" (that is, typically recurrent) usage of Rite II in the 1979 *Book of Common Prayer*. This reflection follows the usual distinction between Liturgy of the Word and Liturgy of the Sacrament, supplying along the way traditional Latin or Greek titles for key stages of the liturgical action. I have not tried to discuss here other features of common worship that could be profitably addressed in a longer treatment: the Prayers of the People, for example, the Penitential Order, the sermon, the Lord's Prayer, hymnody, or various seasonal observances within the Church Year.

Liturgy of the Word

Kyrie eleison, Christe eleison, Kyrie eleison

It is hard to ignore the penetrating simplicity of this ancient litany: "Lord, have mercy. Christ, have mercy. Lord, have mercy." As rendered above in the ancient Greek, it is the only non-English language portion of the eucharistic liturgy officially authorized for optional use in the Episcopal Church. For many of us, the elemental force of its supplication becomes all the more striking as rendered musically through the many evocative Mass settings composed by Bach, Mozart, Beethoven, and others.

The litany's disarmingly direct invocation of "mercy," reinforced through repetition with clear reference to the Trinity, gives voice to our visceral sense of an unspecified yet immense neediness and vulnerability. The neediness is most immediately ours, as human worshippers aware of our own failures and sin, combined with life's further diminishments in the form of loss, pain, and death. To all this, the liturgy's first response is a simple, unelaborated *cri de coeur*: "Lord have Mercy."

Yet the supplication voiced here also surpasses our personal and human wants. It is, in fact, all-encompassing. It voices that sense of yearning and imperfection we understand to be shared not only by all human souls, but applicable as well to every being throughout the whole of Creation. The point is best reflected in the more recent and accurate English translations[7] circulated for various liturgical settings of the ancient Greek formula, *Kyrie eleison*, phrased not as "Lord, have mercy upon us" (found, for example, in Rite One and versions of the Episcopal Church's *Book of Common Prayer* prior to 1979) but as that core, radically universalized appeal: "Lord, have mercy."

---

7. As noted by Marion J. Hatchett in his *Commentary on the American Prayer Book* (New York: Seabury Press, 1980), 320, the translation in question from Rite Two is one of those offered by the English Language Liturgical Consultation (ICET), an ecumenical body of liturgists.

Mercy, like the divine love we presume to be enfolded within it, thus covers indeed a multitude of sins . . . and afflictions and losses. In its cosmic reach *Kyrie eleison* reminds us, too, that the character of the One addressed in this prayer is not a vengeful deity to be placated but the very source of compassion, grace, forgiveness, and love.

Gloria in excelsis Deo
"Glory to God in the highest, and peace to his people
on earth."

While the Kyrie sounds a clear note of neediness and supplication, from and for us as humans, but also on behalf of all God's creatures, the Gloria then shifts attention to declaring outright the majesty and singularity of a Triune God. It is a hymn focused not on petitions but on an outpouring of praise, worship, and thanksgiving: "We worship you, we give you thanks, we praise you for your glory."

Understandably, this hymn is sung, from the standpoint of its human congregation, in an unmistakably human voice. Lacking any direct reference to the physical world, it can scarcely be described as "ecocentric" or "biocentric." But neither is its perspective best described as human-centered, or "anthropocentric." Above all, the Gloria comes across as emphatically God-centered, or theocentric. And we might well imagine the "peace" it invokes, though settled most explicitly over God's "people on earth," to include that Hebraic *Shalom*, a vision of universal harmony, we'd pray to see extended across the larger life of our world.

Credo in unum Deo
We believe in one God,
the Father, the Almighty,
maker of heaven and earth,
of all that is, seen and unseen.

The *Credo* serves to encapsulate, for the collective assent of worshippers, several key teachings of the Christian faith. Even more pointedly than the Apostles' Creed, this statement, hammered out from debates at fourth-century councils in Nicene and Constantinople, aimed to clarify the Church's understanding of a unique interplay between divine and human natures in the person of Christ. In accord with New Testament writings such as those of St. John and St. Paul, the Creed affirms that not only God the Father but the Son and Holy Spirit have, from the first and ever thereafter, been active agents to the processes of Creation. Through Christ, too, as the Logos of God, "all things were made." No less does the Holy Spirit merit worship as "the Lord, the giver of life."

I think it worth our pausing to reflect on two other features of this Creed with noteworthy bearing on deep green faith. To begin with, it testifies to the wondrous scope and variability of God's creation. It posits a created order encompassing multiple dimensions of reality—spheres of being both material and spiritual, temporal and timeless, including that which lies within, beyond, and beneath our usual sense experience. There is a startling amplitude to the cosmos we inhabit, as underscored by latter-day astronomical science. God, the ultimate source of life, is declared to be maker "of all that is, seen and unseen." The solid, monosyllabic resonance of "all that is," as phrased in the contemporary language, strikes me as especially compelling. For its force to claim recognition, however, those reciting the line would do well to observe the comma's worth of pause before rushing ahead to "seen and unseen"—as I'm afraid happens more often than not.[8] In any case, the reference to things

---

8. Admittedly, this point is not directly applicable to other vernacular translations that use slightly different phrasing and punctuation or, for that matter, to the Creed's originally unpunctuated Greek and Latin forms. Evident in any case, however, and worth taking freshly to heart, is the statement's insistence upon an all-encompassing Creation, distinguishable from its Creator, that surpasses materiality and the reach of our bodily senses.

"unseen" or "invisible" underscores the Creed's expansive sense of creation. Air, wind, heat, love, justice, black holes, gravity, antimatter, quarks, and much of the electromagnetic energy that surrounds us—how *many* unseen things we know to be real surpass our physical senses or our capacity to measure!

Another salient feature of the creed, from the standpoint of green faith, is its insistence upon the integral unity and coherence of the universe where we dwell, which mirrors in turn the inviolable unity of the godhead. The unity-in-diversity of the Lord's creation extends across intergalactic space and across all boundaries of time and place so as to interlink "heaven and earth," things "seen and unseen." And as an immense coalescence of mutually dependent beings, creation as a whole resembles that earthly meta-system of interactive life forms we regard as inherently ecological.

Within the Nicene Creed, this vision of an ultimate and integral unity can be seen in the text's language, evoked especially its adjectival reiteration of the word "one." So we're reminded repeatedly of our ties to a primal oneness—through our shared belief in "one God," in "one Lord, Jesus Christ" who is "one being with the Father," in "one baptism" and "one holy catholic and apostolic church." No wonder, then, that thoughtful engagement with the Creed can help prepare us to appreciate the significance of our sharing thereafter, at the table sacrament, in the one bread and one cup. St. Paul underscores the point that "because there is one bread, we who are many are one body, for we all partake of the one bread" (1 Cor. 10:17). Consider, too, what it meant for Julian of Norwich, that celebrated English mystic from the fourteenth century, to describe the visionary process of our becoming, as she put it, "oned" with God.[9]

---

9. In chapter 18 of *Revelations of Divine Love* [adapted by Dom Roger Huddleston, OSB (Mineola, New York: Dover, 2006)] Julian writes, for example, of envisioning a "great oneing betwixt Christ and us," and of how "Christ and she [our Lady, Saint Mary] were so oned in love that the greatness of her loving was cause of the greatness of her pain" (34–35).

The Creed's closing statement whereby "we look for the resurrection of the dead, and the life of the world to come" recalls yet another aspect of unity-in-faith. For God's ultimate renewal of "all that is, seen and unseen," in the course of New Creation, is to be understood as a transformative completion—not annihilation—of the beauty, love, and grace already made manifest in Creation.

Liturgy of the Table Sacrament

New Creation and Great Thanksgiving

Though centered inevitably in a particular place, time, and cast of human participants, every eucharistic celebration also reflects some cognizance of its setting within the larger context of Creation, the Lord's grand theatre of "all that is, seen and unseen" throughout the earth and beyond it. This point is expressly recalled, for example, through the remarks included in the Episcopal Church's Eucharistic Prayer B: "We give thanks to you, O God, for the goodness and love which you have made known to us in creation." Rite I, too, at this stage of liturgy recalls once again how God "didst create heaven and earth." Prayer C in Rite II expounds still more plainly upon God's rule over a created order that encompasses "the vast expanse of interstellar space," countless bodies within such space, and "this fragile earth, our island home."[10] It is a theme sounded, too, in Eucharistic Prayer D, which holds ecumenical standing[11] as a liturgical text accepted with minor differences in Roman Catholic, Lutheran, Methodist, and other denominations: "Fountain of life and source of all goodness, you made all things and fill them with your blessing; you created them to rejoice in the splendor of your radiance." Eastern Orthodox theologians such as Alexander Schememann,

---

10. *Book of Common Prayer*, 368, 370.
11. See Hatchett, *Commentary on the American Prayer Book*, 377.

John Zizioulas, and John Chryssavgis have been particularly disposed to advance recognition of the ways in which the eucharistic sacrament embodies, in turn, "the sacramentality of creation itself."[12]

And beyond the now-superseded tangle of Reformation-era and other disputes about the modality of Christ's presence in the Eucharist, the Church's mainstream tradition has long affirmed that a *transformation* of various sorts takes place during this most solemn stage of the liturgical action. The metamorphosis in question surely involves, though is not restricted to, that portion of the eucharistic prayer in which material elements of bread and wine are first consecrated, with benefit of the Holy Spirit's invocation (*epiclesis*), before they are shared within the congregational body. The liturgy's transformational dynamic across time has both immediate and more sweeping eschatological implications toward the future, as suggested by the proclamation of faith that "Christ has died. Christ is risen. Christ will come again." Metamorphosis likewise pertains to the inward hope of personal and collective renewal on the part of communicants, as well as to those more objective, material changes always at play in the course of our growing, preparing, and absorbing foodstuffs integral to the eucharistic feast.

Yet the transformation of elements at the core of eucharist, whether or not described in terms of "transubstantiation," should not be regarded as an obliteration of their materiality as bread and wine. The Eucharist is steeped in sacred mystery, to be sure. But we know the nature of *all* life, spiritually and existentially as well as materially, to be marked by

---

12. Alexander Schmemann, *The Eucharist: Sacrament of the Kingdom* (Crestwood, New York: St. Vladimir's Seminary Press, 1988), 33. Another useful resource, among other relevant commentaries from this tradition, is the essay collection, *Toward an Ecology of Transfiguration: Orthodox Christian Perspectives on Environment, Nature, and Creation*, ed. John Chyssavgis and Bruce V. Foltz (New York: Fordham University Press, 2013).

transformative change. In that regard, it is worth recalling how several stages of formative change, combining earth-grounded processes of growth with those of human agency, contribute to making victuals such as bread and wine, with several more steps of digestion and nutritive assimilation needing to occur in our own bodies when they are consumed. In the sacrament's ingathering of material elements "you may observe," according to seventeenth-century divine Lancelot Andrewes, "a fullness of the seasons of the natural year; of the corn-flour or harvest in the one, bread; of the wine-press or vintage in the other, wine."[13]

Eating, after all, is an emphatically bodily act. Eating and drinking with others, at a shared festal meal, not only solidifies our common humanity but can offer us a taste of the divine—a foretaste, in fact, of the heavenly banquet.[14] For believers, these several traits find supreme expression in the Eucharist. Enacted through the drama of this sacrament is a vision of mystical communion that links Christ's broken yet risen body with the body of worshippers as well as with the vast body of creation and of humanity at large. As Norman Wirzba observes, "By eating at the Lord's Table, people are given here and now a glimpse of heaven as the sort of life God desires for the whole creation."[15] And while we are especially cognizant today, thanks to advances in medical science, of the elusive ambiguities and complications inherent in all "body" language, we cannot ignore that startling insistence on

---

13. Sermon of Lancelot Andrewes, cited in A.M. Allchin, *The Living Presence of the Past: The Dynamic of Christian Tradition* (New York: Seabury Press, 1981), 60–61.

14. For a searching and illuminating exposition on the broadly sacramental, life-encompassing implications of these processes, see Norman Wirzba, *Food and Faith: A Theology of Eating* (New York and Cambridge: Cambridge University Press, 2011), especially chapter 5 on "Eucharistic Table Manners: Eating toward Communion," 144–178.

15. Norman Wirzba, *Food and Faith*, 153.

materiality in Jesus's words, spoken again at the consecration: "This is my body. . . . This is my blood."

The Offertory portal to the Great Thanksgiving calls special attention to the material face of creation, in the form of gifts presented before the altar. This physical oblation of the community's "first fruits," drawn from the earth but refined through human labor, has deep roots in the rituals of vegetative and animal sacrifice practiced in ancient Israel, as well as in primordial rites of other cultures. That something of God's earth we otherwise fancy *we* possess must be surrendered to allow something still greater to emerge, hallowed and transformed—some such life-principle seems to underlie the elemental impulse toward ritual sacrifice. Much adapted and refined, it's also an impulse present in the Offertory prelude to the Great Thanksgiving, as this sequence from the text of present-day Roman Catholic liturgy brings to light:

> Blessed are you, Lord God of all creation, for through your goodness
> we have received the bread we offer you: fruit of the earth and work of human hands, it will become for us the bread of life.

> Blessed are you, Lord God of all creation, for through your goodness we have received the wine we offer you: fruit of the vine and work of human hands, it will become our spiritual drink.[16]

Evelyn Underhill points out that "wild animals and fruits are never used by agricultural peoples for the purposes of sacrifice" because "they must give something into which they have put their own life and work," as a small token of that

---

16. "Text of the English Roman Catholic Mass," CatholicBridge.com, accessed June 22, 2023, https://www.catholicbridge.com/catholic/catholic-mass -full-text.php.

larger self-oblation they intend. Yet only in "the absolute oblation of the Cross," she reminds us, is the "full meaning" of any
such offering disclosed. I take this "absolute oblation" of Christ
to supply the Offertory's motive force. Accordingly, Underhill finds that "Perhaps the most significant development in
human religion has been the movement of the idea of sacrifice from propitiation to love"—a love we can see distinctively
enacted within the communion sacrament as a whole.[17]

> *Sanctus, Sanctus, Sanctus*
> Holy, holy, holy is the Lord of hosts,
> the whole earth is full of his glory.

In its original Hebrew context, this chant emerges from
a key moment in the story of Isaiah's commissioning as a
prophet. Its setting amid the ritual worship taking place in the
Temple of Solomon is fraught with signs of numinous power.
Vivid appeals to several bodily senses permeate the scene. Isaiah envisions the Kingly Lord of the universe enthroned on a
"high and lofty throne," bearing an immense robe that "filled
the temple" as he is surrounded by a bevy of winged, celestial
creatures chanting praises to God's glory. The smoke and odor
of incense fills the place, while the repeated cry of a superla-
tive "holy, holy, holy" from seraphs testifies to the incursion of
a transcendent presence, a touch of heaven itself. And Isaiah,
more directly than Moses had before him, claims that on this
occasion he actually "saw the Lord" (Isaiah 6:1).

Although the locus of this theophany is centered for the
moment in the Jerusalem Temple on Mount Zion, that house
of the Lord evidently stands as a *microcosm* of the *macrocos-
mic world*, or larger cosmos, that all creatures inhabit.[18] Isaiah

---

17. Evelyn Underhill, *Worship: Man's Response to the Eternal* (New York:
Harper & Brothers, 1936), 50, 52.

18. This commonly recognized correspondence is discussed, for example, by
R. E. Clements in *God and Temple* (Oxford: Basil Blackwell, 1965), 65, 67.

here perceives "the whole earth" indeed to be "full of his glory" (Isaiah 6:3). For us today, roughly a century after astronomer Edwin Hubble's breakthrough findings, this *Sanctus* invocation calls to mind the unspeakably vast expanse of the created universe, which encompasses billions of galaxies, adorned with stars and exoplanets too plentiful to number. Isaiah's all-expansive vision includes, too, the otherwise unseen abode of angelic creatures together with the Kingly Lord of it all, whose "glory," or visible effluence of holiness, engulfs everything. It's an occasion, as Eastern Orthodox liturgy particularly aims to re-enact, wherein one might witness conjoined all time and space, matter and energy, terror fused with transcendent beauty.

Within the setting of Christian liturgy, the *Sanctus* declaration also assumes a slightly different character from that supposed within the original context of Israel's temple worship. Isaiah may well have been a temple priest. And whether we suppose him to have been physically situated squarely in the building's main sanctuary, or possibly even within sight of an open door to the Holy of Holies at the time of his vision,[19] we find no sign of participation in his worship experience on the part of lay devotees, whose own observances would ordinarily have taken place on the outer vestibule or porch. But in most versions of the Christian rite, lay members do congregate of course in or near the main worship sanctuary, where they are expected to join in reciting, intoning, or absorbing a choral rendering of the Sanctus. Liturgical churches regard the site of such observance as a sacred space comparable to Jerusalem's

---

19. Some such theory has been proposed by H. G. M. Williamson, "Temple and Worship in Isaiah 6," in *Temple and Worship in Biblical Israel*, ed. John Day (London and New York: T&T Clark, 2005), 137. Margaret Baker, in her *Temple Mysticism: An Introduction* (London: SPCK, 2013), 2–3 suggests instead that the prophet's vision of the Lord enthroned must be situated in the holy of holies— whether or not Isaiah was literally there," since otherwise the high priest alone, on the Day of Atonement, could be expected to enter that space.

temple. All Christian faith communities, particularly those shaped by Reformed teachings, have also seen the ancient temple to be a type of the human heart, and of the gathered community of faith. In effect, then, the Eucharist's new-order reprise of the temple experience attributed to Isaiah supposes the locus of temple worship to lie at least partly now within worshippers themselves, even as its impression of divine glory also suffuses "the whole earth."

> *Agnus Dei*
> Lamb of God, that takest away the sins of the world, have mercy upon us.
> Lamb of God, that takest away the sins of the world, have mercy upon us.
> Lamb of God, that takest away the sins of the world, grant us thy peace.

Figurative titles for God, or for God-in-Christ, abound throughout texts of our scriptural heritage. The Holy One is variously portrayed as a rock, fortress, redeemer, Father, Creator, Son of Man, Son of God, or shepherd, in addition to other attributions. Among these titles, the identification of Christ as "the Lamb of God" certainly has a familiar ring, having been set before us in scriptural references such as those in John 1:29 or Revelation 5:6 and then kept alive for centuries through liturgy and iconographic renderings. It is so familiar, in fact, that we are apt to miss the rare, arresting force of imaging Christ here as a nonhuman animal.

This provocative identification, like C. S. Lewis's audacious portrayal of the Son as a wild beast, the lion Aslan, in his *Chronicles of Narnia*, thus presses us to reflect not only on Christ's humanity but on his still broader and necessary participation in the whole of creaturely animality. Of course the species indicated, a lamb—one evidently slated for slaughter—has sacrificial meaning. Its meaning in the *Agnus Dei* enlarges upon that of another great feast, featuring the Passover Lamb. Some of this sacrificial legacy, which eventually came to be associated with a "payback," substitutionary theory of atonement, must be considered problematic. Arguably, though, the form of sacrifice dramatized in the *Agnus Dei* sequence is not propitiatory, meant to placate a distant and angry Lord, so much as an expression of God's loving, self-emptying gift of salvation to God's own creatures, bestowing upon them the fullness of mercy, peace, and forgiveness.

For several reasons, then, this traditional spot of liturgy, though less commonly prayed today than formerly, plays a noteworthy role within the eucharistic feast. Moreover, its re-iterated supplication for "mercy" toward the liturgy's close offers a fitting bookend to the earlier introduction of that theme in the Kyrie supplication.

The Blessing and Dismissal
Let us go forth into the world,
rejoicing in the power of the Spirit.
Thanks be to God.

This brief coda to the formal liturgy is a telling point of transition. It marks, of course, a necessary shift in the faith community's focus of attention: from one place, in the designated worship space, to another, somewhere in the quotidian secular world. But we should not understand it to mark, as might be supposed, a drastic shift of spiritual orientation. It is not as though we have, during the worship interval, merely absented ourselves for a time from God's created order in order to recharge our batteries for the real business of working, and of working out our salvation, in quite another realm—that is, in the "real" world we imagine to be our existential home. What the dismissal means to signify, in other words, is more of a continuity, an extension, an effectual widening of spiritual consciousness, than a clear break or endpoint. It marks a shift of locus but not of essence. Having for a time *worshipped* eucharistically, congregational members are called to "keep the feast" by *living* eucharistically—that is, by continuing in diverse ways to render praise and thanksgiving to God, in joyous consort with all creatures, and to continue offering the Creation back to the Creator.

The "world" into which congregants are called to "go forth," according to this particular dismissal option in Rite II, cannot be construed, therefore, as a realm distinct from Creation—or distinct from the liturgy itself, since "the world, to which renewal is promised, is present in the whole eucharistic celebration."[20] And it is certainly not removed from planet earth. To be sure, the equivalent to our English "world" can, in biblical usage, variously connote the planet, all its human inhabitants, the social order, an ungodly sector of the social order, or the entire cosmos. But in the sendoff context of this dismissal, the "world" is best understood as just another, relatively secular face of Creation situated beyond church walls. In one creaturely form or another, those elemental eucharistic activities of eating and drinking take place there as well, unceasingly and universally. God's joyous desire "to bless his Creation and to bring it to its fulfillment"[21] is perpetually sustained there, too. And if, finally, we are to fulfill the Eucharist's intended reach and significance beyond church walls, we will, in going forth, carry into the world that sacramental vision of all life it so palpably embodies.

---

20. *Baptism, Eucharist and Ministry*, 14.
21. Graham Tomlin, *The Widening Circle: Priesthood as God's Way of Blessing the World* (London: SPCK, 2014), 6. Tomlin comments further in his study on how he sees humanity "called to play a priestly role between God and Creation" (12).

Conclusions

Developing fuller cognizance of how our usual practices of eucharistic worship both embody and foster this sacramental vision of all life, an endeavor toward which I hope to have contributed something here, strikes me as the most promising approach to reconciling liturgical worship with the "green" imperative of contemporary Christian faith. Such a Creation-infused vision is not solely the product of present-day thinking, however. We may be in a better position today than in previous decades to understand the ways in which love and care for the earth are a central, rather than supplementary or special-interest, component of Christian faith and practice. Yet affirming the larger, cosmic import of the eucharistic feast qualifies, to recall St. Augustine, as a project at once "so ancient and so new."

There is thus longstanding precedent, particularly in the Eastern Church, for identifying the "communion" enacted in the eucharistic feast not only with an individual believer's communion with God in Christ but also with the inherently ecological principle of interchange *among* human worshippers and with every other member of God's created order on earth and throughout the heavens. Communion with Creation is part of this equation. There is also ancient precedent, in various of our liturgical traditions as well as in the Psalms and other biblical texts, for enlarging our appreciation of how human worship can give voice and language to otherwise voiceless nonhumans, affirming our solidarity with those others—especially fellow mortals and sentient beings—inhabiting God's broader community of creation.

There is likewise ample precedent, from well before our own day, for savoring the material earthiness of the eucharistic feast in whatever liturgical mode it might be celebrated. As I have already noted, in the seventeenth century Lancelot Andrewes was already cognizant of how the sacrament's integration of material elements expressed "a fullness of the seasons of the natural year; of the corn-flour or harvest in the one, bread; of the wine-press or vintage in the other, wine." Additionally, Evelyn Underhill, while writing in 1940 with reference mainly to the Lord's Prayer, called attention to the earthy dynamic of nutritive processes also at play within the Eucharistic action:

> The symbolism of food plays a large part in all religions, and especially in Christianity. As within the mysteries of the created order we must all take food and give food—more, must take life and give life—we are here already in touch with the "life giving and terrible mysteries of Christ," who indwells that order; for all is the sacramental expression of His all-demanding and all-giving Life. . . . Starvation both of body and soul is an ever-present possibility. "Thou feedest thy poor ones abundantly with heavenly loaves!" says an ancient prayer of the Spanish Church; a declaration beginning in man's Eucharistic experience, which spreads to embrace that primal Charity by which the cosmos is sustained.[22]

In a few church settings today where "green team" members aspire to heighten environmental consciousness by attending in some detail to how worship is conducted, all sorts of practical questions can arise, such as what sort of eucharistic bread and wine should be supplied, how one should phrase the intercessory prayers, whether or how to provide altar flowers—or, for that matter, how best to handle the distribution of refreshments at the post-worship coffee hour. These questions can be worth pondering and addressing. But they should never, in my view, distract attention from the central mystery of faith

---

22. Evelyn Underhill, *ABBA: Meditations Based on the Lord's Prayer* (London: Longmans, Green and Co., 1940), 52, 56.

represented in the eucharistic feast itself. In that regard, the sacrament's "green" essence is to offer the world an expansive prospect of communion that not only draws together all sorts and conditions of humankind, living and dead, but also celebrates and offers back to God the fullness of Creation in all its forms, reaching across every species divide.

Questions for Discussion and Reflection

1. According to the Faith and Order Commission's landmark ecumenical statement from 1982, the Eucharist's "central act of the Church's worship" expresses "great thanksgiving to the Father for everything accomplished in creation, redemption and sanctification" and is "the great sacrifice of praise by which the Church speaks on behalf of the whole creation." What implications does this statement bear for the Church's understanding of green faith? And since the Commission's statement reflects an agreement reached only by theologians and church leaders from the various denominations, how can its import be known and embraced more widely by ordinary lay people and clergy?

2. In tandem with existing eucharistic texts, liturgical churches allow for considerable latitude in hymnody and authorized forms of intercessory prayer. What specific selections in this vein might you recommend that honor green faith but that also develop suitable seasonal themes and maintain high standards of theological and literary expression?

3. What strategies could best be incorporated into eucharistic preaching so as to expand awareness of the beauty and fragility of creation without leading worshippers to dismiss green faith as a "special interest" or partisan topic?

4. What non-eucharistic patterns of green liturgy could best be developed for discrete occasions, including ones where an ecumenical or interfaith approach is expected?

Afterword

We began this inquiry holding in our minds two distinct photo images of our earth. One of these, taken from very far away by the Voyager 1 spacecraft, pictures our planet as an inconspicuous blue dot amid an oceanic blackness more vast than had been imagined throughout most of human history. Its impression of earth's physical standing in the universe is sobering. The other photo, snapped by astronaut Bill Anders from moon orbit during the 1968 Apollo 8 mission, is much more reassuring, even inspiring. This iconic Earthrise image portrays our home place as a lovely abode apparently replete with liquid water, solid landforms, a stable and temperate atmosphere, and God's own plenty of life-sustaining potential. Viewed from such a novel standpoint near the moon, earth looks to be a veritable Paradise, exactly the sort of planet we would dream to find and inhabit if we didn't already live here. Yet it is worth pondering how *both* of these images—the pale blue dot as well as the bountiful blue orb—dramatize principles crucial to grasping the theology and practice of earthcare needed in these times.

That blue dot calls us to realize a level of collective humility about our place in the larger scheme of things that our species has rarely been willing or able to absorb. Within this setting, our home strikes the eye as distressingly minute, fragile, and inconsequential in its physical scale—but also as far from central in its galactic or cosmic positioning. It's a put-down consonant with the mood of God's withering challenges to Job, when God speaks out of the whirlwind: "Where were you when I laid the foundations of the earth? Tell me, if you have understanding" (Job 38:6).

Seeing ourselves in such a diminished light is chastening, even shocking. But it also awakens us to our limitations, to the fragility and transience of our planetary civilization. It should stir us, as Carl Sagan once observed, "to preserve and cherish the pale blue dot" that is "the only home we've ever known."[1] It just might induce us to ward off catastrophe in the face of forces— including those driving climate change—larger than ourselves though it is we ourselves who loosed them on the world.

Set against this starker image, the Earthrise photo lets us envision, more graphically than ever before, the singular beauty, stability, and fecundity of our home planet. Its more hopeful face of our place and destiny is also essential, for it helps us recall a creation expressly conceived by its Creator to be "very good." We have every reason to marvel not only at the grandeur and blessedness of this earth but that it exists at all. Every reason to wonder at how its gratuitous presence should so unpredictably enfold and sustain us. The Psalmist's songs resound with praise and thanksgiving at the sheer *giftedness* of this creation.

All is not well beneath the outward view of earth's loveliness, we must recall. Yet unlike every other planet in our solar system, this one might well inspire someone gazing on it to exclaim, after novelist John Cheever, "Oh What a Paradise it Seems."

So indeed, a paradise it *seems*, though for Earth to *be* that paradise requires ceaseless care, and the exercise of an active imagination befitting our role as co-creators with God. Surely it is no coincidence that the biblical ideal of our participation in a harmoniously adjusted, biodiverse life on earth typically takes the shape of a garden paradise. That vision of God's *shalom* is pictured both as our primal abode, in the Edenic overture to the Book of Genesis, and as the consummate garden, bearing

---

1. Carl Sagan, *Pale Blue Dot: A Vision of the Human Future in Space*, 8.

the ultimate fruits of God's redemption, in Scripture's closing Book of Revelation. Revelation's ultimate garden, epitomizing the new heaven and earth, is said to encompass the holy city of God as well as the river and the tree of all life. As such, it not only recalls but exceeds the blessedness of Eden's original paradise, through the redemptive work of One honored as the story's Alpha and Omega, the beginning and the end.

One picture that for me tallies well with this poetic imagery flows from my participation years ago in a Christian fellowship, called the Hie Hill Community, that met and prayed regularly at a rural property in Westbrook, Connecticut, near Long Island Sound. The presiding spirit of this group was Olive, an energetic woman much devoted to tending and keeping the property's impressive array of floral and vegetable plantings. Gardening seemed lodged in her being. From time to time she also recruited all of us, fellow members of her community, to this labor. Olive especially liked to remind us how, according to the Gospel of John, Mary Magdalene had, on entering Jesus' tomb and encountering there the risen Christ, first supposed him to be the gardener.

Well, why not? Olive felt this to be one of those mistaken impressions that cloak subtler truths. Like the artists before her who had likewise been drawn to reimagine that scene, she thought it quite fitting that Jesus should have been seen as a gardener, the One commissioned from eternity to oversee the planting and keeping of all things. For me that offers a heartening, down-to-earth reminder of what, even or especially in this crisis moment, our vocation as creative keepers and friends of this planet might look like in practice. Olive died several years ago, but I like to think of her as still working diligently there, trowel in hand, steeped in leaves and dark soil, under the approving eye of her master gardener. Hers—and ours as well.